THE ENCYCLOPAEDIA OF IGNORANCE

VOLUME 1

PHYSICAL SCIENCES

Also available from Pergamon Press in a flexicover edition
The Encyclopaedia of Ignorance — Life Sciences and Earth Sciences

THE ENCYCLOPAEDIA OF IGNORANCE

Volume 1

PHYSICAL SCIENCES

Edited by

RONALD DUNCAN

and

MIRANDA WESTON–SMITH

PERGAMON PRESS

OXFORD · NEW YORK · TORONTO · SYDNEY · PARIS · FRANKFURT

U.K.	Pergamon Press Ltd., Headington Hill Hall, Oxford OX3 0BW, England
U.S.A.	Pergamon Press Inc., Maxwell House, Fairview Park, Elmsford, New York 10523, U.S.A.
CANADA	Pergamon of Canada Ltd., 75 The East Mall, Toronto, Ontario, Canada
AUSTRALIA	Pergamon Press (Aust.) Pty. Ltd., 19a Boundary Street, Rushcutters Bay, N.S.W. 2011, Australia
FRANCE	Pergamon Press SARL, 24 rue des Ecoles, 75240 Paris, Cedex 05, France
FEDERAL REPUBLIC OF GERMANY	Pergamon Press GmbH, 6242 Kronberg-Taunus, Pferdstrasse 1, Federal Republic of Germany

First edition 1977

British Library Cataloguing in Publication Data

The encyclopaedia of ignorance.
1. Science
I. Duncan, Ronald II. Weston-Smith, Miranda
500   Q158.5   77-30376
ISBN 0-08-021238-7
ISBN 0-08-021230-1 Pbk (Vol. 1)
ISBN 0-08-021231-X Pbk (Vol. 2)

Printed in Great Britain by William Clowes & Sons, Limited London, Beccles and Colchester

CONTENTS

EDITORIAL PREFACE

Compared to the pond of knowledge, our ignorance remains atlantic. Indeed the horizon of the unknown recedes as we approach it. The usual encyclopaedia states what we know. This one contains papers on what we do not know, on matters which lie on the edge of knowledge.

In editing this work we have invited scientists to state what it is they would most like to know, that is, where their curiosity is presently focused. We found that this approach appealed to them. The more eminent they were, the more ready to run to us with their ignorance.

As the various disciplines have become increasingly specialised, they have tended to invent a language, or as we found in the computer field, a jargon almost incomprehensible to anybody outside that subject. We have tried to curtail this parochialism and have aimed this book at the informed layman, though possibly at university level, in the hope that he will be encouraged to read papers outside his own subject.

Clearly, before any problem can be solved, it has to be articulated. It is possible that one or two of our papers might direct research or stimulate it. In so far as it succeeds in stating what is unknown the volume will be of use to science historians. A decade hence many of the problems mentioned in these pages will have been solved.

It could be said that science has to date advanced largely on the elbows and knees of technology. Even the concept of relativity depended on technology to prove its validity. In some disciplines we have already reached the point when the Heisenberg principle applies and the observer alters the object observed. And it may well be in cosmology especially, in our attitudes to space and time, that our concepts are our limiting factor. Perhaps imagination is a part of our technology? Perhaps some answers depend only on asking the correct question?

Ronald Duncan
Miranda Weston-Smith

ACKNOWLEDGEMENT

The Editors wish to acknowledge with gratitude the considerable help and the advice they have received from Professor O. R. Frisch, F.R.S. while compiling this book.

Photograph Bertl Gaye, Cambridge, U.K.

Otto R. Frisch, O.B.E., F.R.S.

Jacksonian Professor of Natural Philosophy, University of Cambridge, 1947-72, now Professor Emeritus.

He is best known as the physicist who in 1939 collaborated with Lise Meitner to produce the first definite identification and explanation of the phenomenon of "nuclear fission", a phrase which he himself coined.

Carried out research in many parts of the world, including Berlin, Copenhagen, Oxford, Los Alamos, Harwell and Cambridge, and is the author of many publications on atomic and nuclear physics.

WHY

Some 15 years ago, WHY was a magic word, used by a small boy to keep Daddy talking.

"Daddy, why does the sun go down in the West?"

"Because West is what we call the place where the sun goes down."

"But why does it go down?"

"It doesn't really; it's the Earth that turns round."

"Why?"

"Because there is no friction to stop it."

"Why . . ."

But what do we mean when we say WHY? We expect some answer; what kind of answer? "Why did Jones break his leg?"

"Because his tibia hit the kerb" says the surgeon.

"Because some fool dropped a banana skin" says Mrs Jones.

"Because he never looks where he goes" says a colleague.

"Because he subconsciously wanted a holiday" says a psychiatrist.

For any event there are several styles of answering the question why it happened. (For further confusion, see Schopenhauer's essay "Die vierfache Wurzel des Satzes vom Grunde".) But when we ask why something is so, then we are on different ground. It would be defeat for a scientist to accept the answer "because God made it so". But often there is another answer which perhaps comes to the same thing. To the question "why do intelligent beings exist?" it seems legitimate to reply "because otherwise there would be nobody to ask that question". Many popular WHYs can be answered in that manner.

Let us try another question. In my car, why does the spark plug ignite the mixture at a particular instant? There are two answers:

(1) because the cam shaft causes a spark at just that instant;

(2) because a spark at that instant gives good engine efficiency.

Answer (1) is what a physicist expects. Still, he may ask why the cam shaft has been made to cause a spark at just that time; then (2) is the answer required. It introduces a new character: the designer, with intelligence and a purpose (in this case the design of an efficient car engine).

The teleological explanation, (2), is here certainly the more telling one (except to that mythical personage, the pure scientist). To the question "why is John running?" the reply "in order to catch the bus" is satisfactory; the reply "because his brain is sending the appropriate messages to his leg muscles" (though basically correct) would be regarded as a leg-pull.

A couple of centuries ago, physical laws were often formulated in teleological language; that this was possible appeared to show that the laws had been designed to fulfil some divine purpose. For instance, a beam of light passing through refracting media (as in a telescope) was shown to travel along the path that requires the minimum time, according to the wave theory of light. But the basic law of refraction can be deduced without reference to that parsimonious principle and, moreover, allows light to travel equally on a path that requires not the least but the most time (at least compared to all neighbouring pathways).

Today such minimum (or maximum) principles are regarded merely as pretty (and sometimes useful) consequences of more basic laws, like the law of refraction; it was anyhow always obscure what divine purpose was served by making light go the quickest (or sometimes the slowest) way. Teleological explanations are not accepted nowadays; not in physics.

In biology it is otherwise; nobody doubts that many features of animals serve a purpose; claws serve to kill, legs to run, wings to fly. But is it a divine purpose? The great debate has not altogether ceased, but the large majority of scientists are agreed that natural selection can account for the appearance of purposeful design, even though some of them find it hard to imagine how such a marvellous instrument as the human eye (let alone the human brain) could have developed under the pressure of natural selection alone.

The power of artificial selection is, of course, well known to any plant or animal breeder. Admittedly, natural selection lacks the breeder's guiding hand. But it has acted through millions of years and on uncounted billions of individuals, and its power of favouring any improved adaptation to the life a species has to live is inexorable. The dug-up skeletons of horses show the development, over some millions of generations, of a rabbit-size creature to the powerful runner of today whose ancestors survived pursuit by ever faster predators. Natural selection still works today: of two varieties of moths of the same species, the darker variety predominates in smoky cities where it is well camouflaged, as his lighter cousins are in birchwoods where they in turn predominate.

As to the human eye, any light-sensitive organ, however primitive, is useful, and any improvement in sensitivity, resolution and mobility is strongly favoured by natural selection. But what about feathers? Even if a very unlikely mutation caused a reptile to have offspring with feathers instead of scales, what good would that do, without muscles to move them and a brain rebuilt to control those muscles? We can only guess. But let me mention the electric eel. It used to be a puzzle how his electric organ could have grown to its present size when in its early stages it would have been quite useless as a weapon. We now have an answer: even a feeble electric organ helps with navigation in muddy waters, and its gradual improvement has led, as it were, from a radar to a death ray.

Much about the process of evolution is still unknown; but I have no doubt that natural selection provides the justification for teleological answers.

Finally, let us go back to physics and ask a question to which, it seems, there is no answer: Why did a particular radium nucleus break up at a particular time? When the theory of atomic nuclei was young it was suggested that their complexity provided the answer: an alpha particle could escape only when all the others were in a particular configuration, as unlikely as twenty successive zeros in a game of roulette. Even with the configurations changing about 10^{20} times a second, it could take years before the right one turned up. That theory has been given up; for one thing there are much less complex nuclei with similar long lives.

Probability theory started as a theory of gambling. The apparent caprice of Lady Luck was attributed to our unavoidable ignorance of the exact way a dice was thrown; if we knew the exact way we could predict the outcome. Sure, we would have to know exactly how the dice was thrown and every detail of the surface on which it fell much more accurately than we could conceivably hope to achieve; but "in principle" it would be possible.

From those humble beginnings in a gamblers' den, the theory of probability grew in power until it took over large parts of physics. For instance, the observable behaviour of gases was accounted for by the innumerable random collisions of its molecules. Just like computing the profitability of a gambling house or an insurance company, this could be done without predicting the behaviour of single molecules. It might still be possible in principle to predict where a given molecule would be one second later; but to do that we would have to know the positions and velocities of millions of other molecules with such precision that to write down, not those numbers themselves but merely the number of decimals required, would be more than a man's life work!

With that in mind, you might find it easier to accept that quantum theory uses the concept of probability without justifying it by ignorance. Today most physicists believe that it is impossible even "in principle" to predict when a given radioactive nucleus will break up. Indeed it is only a few properties (such as the wavelength of light sent out by a given type of excited atom) for which the quantum theory allows us to calculate accurate values; in most other cases all we get is a probability that a particular event will take place in a given time.

To some people this idea of probability as a physical attribute of, say, an unstable atom seems distasteful; the idea of inexorable laws, even if we can never follow their work in detail, has not lost its appeal. Einstein felt it was essential; "God does not play dice with the world" he said. Could not the seeming randomness of atomic events result from the activities of smaller, still unknown entities? The random movements of small particles (pollen grains, etc.) in a fluid, observed in 1827 through the microscope of a botanist, Robert Brown, were later understood as resulting from the impact of millions of molecules, whose existence was merely a matter for speculation in 1827. Perhaps we shall similarly explain the random behaviour of atoms, in 40 years or so?

Such entities, under the non-committal name of "hidden variables", have been speculated on; so far they have remained hidden. Should they come out of hiding they would probably do no more than restore the illusion that the behaviour of atomic particles can be predicted "in principle". On the other hand, they may possibly predict new and unexpected physical phenomena, and that would be very exciting. I have no serious hope of that, but I can't foretell the future.

Sir Hermann Bondi, F.R.S.

Chief Scientific Adviser, Ministry of Defence and Professor of Mathematics at King's College, University of London.

Formulated the Steady State Theory of the origin of the universe with Thomas Gold.

Director-General of the European Space Research Organization 1967-71. Former Chairman of the U.K. Ministry of Defence Space Committee and the U.K. National Committee for Astronomy, and Secretary of the Royal Astronomical Society. Published widely on many aspects of Cosmology and Astrophysics.

THE LURE OF COMPLETENESS

Theories are an essential part of science. Following Karl Popper it is clear that theories are necessary for scientific progress and equally clear that they must be sufficiently definite in their forecasts to be empirically falsifiable. However, it does not follow that it is desirable, let alone necessary, for a theory to be comprehensive in the sense of leaving no room for the unknown or at least the undefined. In many instances it is the very task of the theory to describe the common features of a large group of phenomena, their range of variety necessarily stemming from something outside the theory. Consider, for example, the Galilean theory that in the gravitational field all bodies suffer the same acceleration and that in a suitably limited volume of space (e.g. a golf course) this acceleration is everywhere the same. Yet golf balls fly about in a vast variety of motions (even if one abstracts from air resistance) according as to how, when and where they have been hit by golf clubs.

The theory is explicitly designed to omit any statement of how the bodies were set in motion and indeed gains its importance from this, for otherwise it would not have its vital universality. By concentration on *accelerations*, dynamical theories allow for an input from arbitrary initial conditions of position and velocity. Any dynamical theory not doing so would be condemned to have an absurdly limited *applicability*. It is not that a dynamicist would regard initial conditions in any sense as inexplicable, but he would not view it as *his* business to explain them.

This is a very widespread characteristic of many scientific theories. Thus in Maxwellian electrodynamics the forces making charges (or current-carrying conductors) are explicitly outside the theory (provided they are electrically neutral), in hydrodynamics the position and motion of the boundaries are viewed as an external input to the theory; in the theory of the excited levels of atoms the exciting agency is taken as externally given, in General Relativity the equation of state of matter is viewed as outside the purview of the theory, etc.

In all these fields a theory that had no room for something outside itself as an essential input would be uselessly narrow. Is this a universal characteristic of scientific theories?

It may be worthwhile trying to classify the exceptions. On the one hand we have *historical* theories. Any theory of the origin of the solar system, of the origin of life on Earth, of the origin of the Universe is of an exceptional nature in the sense stated above in that it tries to describe an event in some sense *unique*.

Looking first at the problem of the origin of the solar system we do not, as yet, know of any other planetary systems though many astronomers suspect that they may be fairly or very common. Up to now the challenge is therefore to devise a sequence of occurrences by which the event of the origin of our solar system *could* have happened. This has turned out to be an extremely severe test, and it has been possible to disprove a variety of theories by demonstrating that no planetary system could have formed thus. We have reached the stage of having theories of how *a* planetary system could have formed, but not one with the actual properties of *our* planetary system, nor the stage of having several theories, each accounting satisfactorily for the features of our solar system, so that each *could* be an adequate description, but leaving us in ignorance of which it was in fact. However, it is reasonable to expect that before so very long we will have significant empirical evidence on the frequency of occurrence of planetary systems and perhaps on what are common features of such systems. Such discoveries will add statistical arguments to be considered and may do much to reassure one that one is not dealing with a truly singular event. In the case of the origin of life it may take much longer before there is any evidence on its frequency of occurrence, and we must recall Monod's warning on the indications in favour of its uniqueness, or at least its extreme rarity. Lastly the origin and evolution of the Universe are almost by definition without peers, and thus of intrinsic uniqueness.

The applicability of scientific argumentation to unique historical events is debatable. A *description* of what happened is surely the most ambitious that could be aimed for. A theory wider than this (e.g. one allowing for a whole range for the time dependence of Hubble's constant) simply does not serve the purpose of accounting for the properties of *the* Universe. For what can be the meaning of the set of unrealised universes? What was it that selected the model with the actually occurring time dependence of Hubble's constant from all the others?

But even in cosmology this demand for completeness that looks so sensible for a global feature like Hubble's constant dissolves into nonsense for characteristics on a lesser scale. The bewildering variety of our Universe is surely one of its most striking features. Even very large subunits, like galaxies, have a taxonomy of amazing complexity, varying amongst each other widely in size, in shape, in constituents, in clustering. Would one really ask of a theory of the origin and evolution of our Universe that it should result in a catalogue of galaxies with their individual properties, arranged cluster by cluster? This is surely driving the demand for completeness to absurdity. The best one can reasonably aim for is that one's theory of the Universe should provide a background against which galaxies of the kinds we know of can form. We could not wish for and could not imagine a theory of the Universe telling us why the Virgo cluster formed in one area and our local system in another near to it. We need a separate input for this, and since something external to the Universe is meaningless, our only alibi can be randomness, which fortunately is an intrinsic property of matter.

Thus we see that even the theory of a necessarily unique system like the Universe not only cannot, but must not, be complete. Similarly we would view with the utmost suspicion a theory of the origin of the solar system that necessarily led to just the set of planetary orbits, masses and satellites that we actually find.

Another case, different from the historical one, where completeness of description looks attractive at first sight, is the case of the study of overall systems, as in thermodynamics. In a certain sense a system in thermodynamic equilibrium is fully described by a small set of parameters (volume, temperature, entropy, etc.), a set we like to think of as complete. However, the very power and elegance of the thermodynamic appraisal lies in its *essential incompleteness*. Whatever the interactions between the constituent particles, whatever their character, the system's parameters give a valid and most useful description of its state. It is true that this is a description of the overall state rather than of all the detail that goes on in the micro-scale, but this detail is generally not required. The fact that we can say a great deal about such a system without knowing about it in detail is a source of pride rather than of regret at the incompleteness of our knowledge.

Similarly the existence of systems parameters such as linear momentum and angular momentum derives its value precisely because completeness of knowledge of the interior of the system is not required. No understanding whatever is needed of anatomy, physiology, or the properties of leather to establish that one cannot pull oneself up by one's bootstraps. Indeed one can argue that science is only possible because one can say *something* without knowing *everything*. To aim for completeness of knowledge can thus be essentially unsound. It is far more productive to make the best of what one knows, adding to it as means become available.

Yet in some sense the lure of completeness seems to have got hold of some of the greatest minds in physics; Einstein, Eddington, Schrödinger and most recently Heisenberg have aimed for "world equations" giving a complete description of all forces in the form perhaps of a "unified field theory". A vast number of hours and indeed years of the time of these towering intellects have been spent on this enterprise, with the end result (measured as one should measure science, by the lasting influence on others) of precisely zero.

In my view it is by no means fortuitous that all this endeavour was in vain, for I think that to aim at such completeness of description is mistaken in principle.

Science is by its nature inexhaustible. Whenever new technologies become available for experiment and observation, the possibility, indeed the probability exists that something previously not dreamt of is discovered. To look only at extraterrestrial research in the last quarter century the van Allen radiation belts, the solar wind, Mars craters, radio sources, quasars, pulsars, X-ray stars are all in this class, owing their discovery to space probes and satellites of various kinds, to radio telescopes and to new instrumentation for optical telescopes. Most of these discoveries were totally unforeseen in the antecedent picture though some of their aspects later turned out to be compatible with it. To suggest that at any stage of technical progress in experimentation and observation we have reached such a level of completeness that it is worth a major effort to encapsulate this imagined (not to say imaginary) completeness in an accordingly supposedly harmonious mathematical formulation makes little sense to me. Where there are empirical reasons to join together previously separate branches then this is a worthwhile enterprise likely to lead to important insights but where there are no such indications one is probably only indulging in a mathematical game rather than in science, for one is hardly likely to find testable observable consequences of such a purposeless unification. Of course, the fashion was started by General Relativity which unified inertia and gravitation with great success. But this was based on Galileo's observation that all bodies fall equally fast. Theory followed experiment by 300 years, for his experiment established the equality of inertia and passive gravitational mass. What is there to guide us in attempting to unify gravitation with electromagnetism and perhaps with weak or strong nuclear interactions? There are no experiments beyond those involved in General Relativity that are joining any of these fields. Until a new technology enables us to perform such experiments, the unification is virtually bound to be sterile.

The counter-argument to my scepticism has generally been that one should rely for guidance on a supposed concept of "mathematical beauty". Experience indicates that while an individual theoretician may perhaps find such a concept heuristically helpful, it is not one on which different people can agree, in stark contrast to the unanimity with which the yardstick of experimental disproof is accepted. Hence the failure of the work of Einstein and others on unified field theories to be followed up, hence the total waste of all this effort. To my mind, which is perhaps not very appreciative of the significance of mathematical beauty, the whole concept looks meaningless and arbitrary depending as it does on whether somebody invents a concise notation or whether a similarity with a previously established mathematical field can be adduced.

The aim of this article has been to show that our most successful theories in physics are those that explicitly leave room for the *unknown*, while confining this room sufficiently to make the theory empirically disprovable. It does not matter whether this room is created by allowing for arbitrary forces as Newtonian dynamics does, or by allowing for arbitrary equations of state for matter, as General Relativity does, or for arbitrary motions of charges and dipoles, as Maxwell's electrodynamics does. To exclude the unknown wholly as a "unified field theory" or a "world equation" purports to do is pointless and of no scientific significance.

R. A. Lyttleton, F.R.S.

Professor of Theoretical Astronomy, and Fellow of St John's College, University of Cambridge.

Gold Medallist of the Royal Astronomical Society 1959, and Royal Medallist of the Royal Society 1965. Member of the Institute of Astronomy, Cambridge.

Research interests include astrophysics, cosmogony, physics, dynamics, and geophysics.

THE NATURE OF KNOWLEDGE

If most of us are ashamed of shabby clothes and shoddy furniture, let us be more ashamed of shabby ideas and shoddy philosophies.

Albert Einstein

When asked where one would like to see scientific research directed, there springs to mind the question whether truly scientific research can be directed at all. It is true that once the principles of a subject have been laid down for us by geniuses such as Newton and Maxwell,[†] then research could be directed to studying their consequences, though even here great ingenuity requiring a high order of mathematical skill and imagination to overcome difficulties may be needed to pass from the principles to accounting for known observations and also making future predictions. The theory of the motion of the Moon provides a case in point: deep problems of mathematics have arisen all along and have had to be surmounted with the outcome of increasing accuracy of prediction, though even today, three centuries after Newton, there still remain puzzling features of the lunar motion. Are these difficulties of importance, and how much effort should be devoted to understanding them sufficiently to tackle them with success? Who is to decide such questions, and how can those doing so come to a proper assessment of such matters? With the early work to achieve more accurate measures of atomic weights, considered to be necessarily in integer-ratios, one can imagine proposals for such research turned down as of unlikely value, yet hidden there was the superlative power of nuclear energy.

Before stating my view as to where I consider effort might best be put, let us notice that our educational systems provide instruction from infancy up to doctoral level on everything appertaining to science, mathematics pure and applied, chemistry, physics, and so on, in order that pupils may achieve familiarity and facility with a subject as it exists. But the ideal objective, if a subject is not to become moribund, should be to learn it so well as to be scientifically critical of it with the object of bringing about advances in it. Sums of money vaster than ever before are enabling huge numbers of people to be drafted into scientific research, with all concerned inspired by an earnest hope that important contributions in the shape of new discoveries and fundamental principles will be made thereby. But when it comes to novel ideas, to the proposal of new hypotheses for realms in which there may be insufficient or even no established principles at all to serve as a guide, then controversy can arise and acute differences of opinion emerge, not always expressed in terms that pay due regard to the amenities of proper scientific discussion. Indeed, so strongly may some believe in their ideas, and so desirable may they consider it to promulgate them widely, that resort may be made to inadmissible ways and means to shield and defend them. (If you, dear reader, have never had any experience of this sort of thing, then your case has been more fortunate than most.)

This having briefly been said, my answer would be that the prime need of science today, so urgent that the body-scientific may choke to death in some fields of endeavour if nothing is done, is for steps to be taken to inform all those working in science what Science is really about, what is its true objective. Of course every scientist thinks he knows this, and this is right – he thinks he does, yet it has to be admitted that many very serious-minded, solid, and knowledgeable people work hard in science all their lives and produce nothing of the smallest importance, while others, few by comparison and perhaps seemingly carefree and not highly erudite, exhibit a serendipity of mind that enables them to have valuable ideas in any subject they may choose to take up. Few of the former have ever stopped off from their industrious

† My examples have to be from fields I have studied, and are not meant to suggest that there have not been equal men in other fields.

habits to consider the question of How does one know when one knows?! It is true that much has been written discussing the subject, but mainly by writers handicapped by lack of experienced appreciation of the technicalities of mathematics and science, and even lack of acquaintance with the very things they are talking about. It is reported that a somewhat pretentious hostess once asked Einstein what were the philosophical and religious implications of his theory of relativity, to which he replied, of course correctly, "None as far as I know". Occasionally a Karl Pearson may come along, leaving aside his regular work, to set down his considered views of the matter for the intended benefit of others, but praiseworthy as such excursions may be, they may do more harm than good if they misconceive the matter, as indeed Pearson did. The few that properly understand the nature and object of science are usually as a result so busy with the work created by their attitude, which leads them to unending fruitful research, that they have little time for anything but actually *doing* science. Just as, contrariwise, those that do not understand the real object of science gradually come to rest in it and perhaps take up administration wherein they may be a ready prey to non-scientific approaches, or they may go in for popular exposition and thereby attain great repute as scientists of eminence by writing for the nursery. The late H. F. Baker, a pure mathematician for whom no point was too fine, when asked his opinion of one of Sir James Jeans's famous popularisations of astronomy, said, "I wish I really knew what he puts on one page".

As best one can apprehend something essentially difficult to discover, the attitude of most of those attempting scientific research seems to be that they believe themselves to be trying to find out the properties of some real material world that actually has independent existence and works in some discoverable structured mechanical way that can be ascertained if only it is studied carefully enough, "Observe, and observe, and observe", and that the "explanation" will emerge of its own accord if only the matter is studied sufficiently long and laboriously. But it will seldom if ever do so: indeed too much in the way of observations can reveal so many seeming complications and contradictions that almost any hypothesis may appear to be ruled out. A new idea may be likened to a new-born babe: it is to be carefully nurtured and given every consideration rather than attacked with the choking diet of a multitude of so-called facts because it cannot prove at once that it will one day grow into a Samson. In this connection, it has been conjectured that had Newton known all the complex details of the lunar motion now known he might never have believed that so simple a law as the inverse-square could possibly explain them. Indeed, even after his death, the observed advance of the perigee (which Newton had in fact already solved) seemed so puzzling that it led to proposals that an inverse-cube term should be added to resolve the problem. It must also be recognised that it is often only when a theory of some phenomenon exists can new discoveries be made and new or improved theories be invented. The advance of the perihelion of Mercury could not even have been conceived or discovered till Newtonian theory was available, while the problem posed by what seems so simple a phenomenon resisted solution for over half a century awaiting the right interpretation, as many consider the general-relativity explanation to be.

But, important as it is to examine and if possible measure every possible aspect of the phenomenon of interest, it may come as a surprise to some and perhaps even be received coldly if it is stated that such activity is only a preliminary to science and not science itself, any more than the manufacture of golf-courses and equipment is golf itself, to give a trifling parallel. The true purpose of science is to invent hypotheses upon which can be developed mathematical theories and formalisms that enable predictions to be made in response to recognised objectives. But the statement raises further questions. First as to where the objectives come from, which is a subjective matter that need not be discussed here if it is admitted that men do find themselves with desired objectives, and second, which will be discussed, is where do the hypotheses come from and how are they to be invented. Newtonian dynamics enables predictions to be made of the motions of the planets and in realms far beyond that, while Maxwell's equations do this for the domain of electricity and magnetism, and quantum-theory for the infinitesimal realm of atoms and molecules.

Now although it is an essential precursor for the formulation of scientific theories that the phenomena of interest are first of all observed and if possible experimented with to at least some extent, it is of the utmost importance to recognise that no secure meaning or interpretation can be given to any observations until they are understood theoretically, or at best in terms of some hypothesis and theory based on it whether right or wrong. And it is here that one comes up against the strangely paradoxical nature of science, for the observations of phenomena are first needed to inspire someone to imagine an appropriate theory, yet they (the observations) cannot be claimed to be properly understood until a formal theory of them is available, and especially is this so where new phenomena are concerned when there are no (theoretical) means whereby the relevance of observations or experiments can be safely assessed. There can be no "facts", no reliable "evidence", until there are hypotheses and theories to test out. Before the advent of gravitational theory, comets were firmly believed to be by far the most important of all heavenly objects, and the official (verbal) theory of the phenomenon was that they were immaterial aethereal portents sent by the gods to warn man of coming violence and pestilence. The railway-lines can be observed definitely to meet in the distance, and interpretation of the observation requires some theory of space, and the lines might even actually meet if some impish engineer decided to make them do so a mile or so away! But the prediction on either hypothesis, parallel or meeting, could be tested out, for example by walking along the track to find out, or making accurate local measurements suggested by either hypothesis and a theory of space (Euclidean in this case would be adequate). The great Newton himself fell into error in this way, for when he found that the theoretical extrapolation of gravity to the distance of the Moon, which necessarily involved the "known" size of the Earth, gave an acceleration inadequate to account for the lunar motion, he abandoned his theory as contradicted by "facts". Six whole years were to go by, with the world the poorer, before it was discovered that the measured size of the Earth was considerably in error and not the gravitational theory. Observations are by no means always to be trusted as reliable guides.

But many "theories" have been constructed and proffered as scientific theories that do not really qualify as such at all. Some amount to no more than a more or less ingenious narration redescribing, sometimes imperfectly and all too often selectively, the data as known, without adding anything to these data, themselves describable in terms of some established theory: the shape of the Earth, for instance, requires for its description the theory of the Euclidean geometry of space, though the hypothesis of its near-sphericity was not always examined by some with due scientific consideration. But unless such a theory can make some prediction of its own, or suggest some crucial experiment, itself a kind of prediction, and few if any verbal theories can do this, it is not a scientific theory in a proper sense. Yet much science-literature today abounds with verbal "theories" wrapped as established stories round data susceptible to more than one interpretation than that proposed and omitting to include equally established data as irrelevant when this is not yet known. Chaucer knew about such verbal rationalisations when he wrote warningly, "You can by argument make a place A mile broad of twenty foot of space". Such theories can add nothing to the data and may even degrade them, and unless the story makes some verifiable prediction unique to the theory itself and does not conflict with other available data, one might just as well accept the data in their entirety. By means of verbal theory, the Moon could be held to the Earth with a piece of string if gravitation were to cease: the force is in the right direction, it would be possible to make a quarter of a million miles of string, and an astronaut could attach one end to the Moon, and so on. But when numbers (theory) are put in, the verbal theory collapses at once, and it is found that a cable of stoutest steel several hundred kilometres thick would be needed. Verbal descriptive theories are analogous to putting a curve through points (representing, say, a series of observations): given a number of points in a plane, say, a curve can be put through them with as great precision as one wishes mathematically, but unless the resulting curve can predict the next point and the next (not necessarily absolutely accurately), the curve is not a theory of the phenomenon and is valueless. The correct theory may not quite go through any of the points.

Before discussing how new ideas and new hypotheses, on which new theories may be built and tested, come to be invented, let us leave the theme for a moment and consider what attitude a scientist should adopt towards such novelties, or indeed towards existing ideas and theories. In a recent lecture Medawar dealt with this briefly by the piece of advice, "Never fall in love with your hypothesis". But controlled energy and enthusiasm are needed to work upon and examine a hypothesis sufficiently carefully, and these are qualities turned on as it were by emotional drive, and if one succeeds in not actually falling in love with one's ideas, which state notoriously weakens if not altogether disables a person's judgement and critical faculty, then how far should one go in relation to a new idea, whether one's own or someone else's? This is obviously a subjective question, but knowably or not, if an idea comes to the awareness of a scientist, he will begin to adopt some attitude to it. This will result from interaction of the idea with all his previous experience, remembered or not, and his character and temperament and so on, and these will combine of their own accord to determine an attitude.

The scientific attitude to adopt in regard to any hypothesis in my view (and we are talking of subjective things) can be represented schematically by means of a simple model of a bead that can be moved on a short length of horizontal wire (see diagram on next page). Suppose the left-hand end denoted by 0 (zero) and the right-hand end by 1 (unity), and let 0 correspond to complete disbelief unqualified, and the right-hand end 1 to absolute certain belief in the hypothesis. Now the principle of practice that I would urge on all intending scientists in regard to any and every hypothesis is:

Never let your bead ever quite reach the position 0 or 1.

This is quite possible, for however close to the end one may have set it, there are still an infinite number of points to move the bead to in either direction in the light of new data or new arguments or whatever. If genuine scientific data reach your attention that increase your confidence in the hypothesis, then move your bead suitably towards 1, but never let it quite get there. If decreasing confidence is engendered by genuine data, then let your bead move towards 0, but again never let it quite reach there. Your changing confidence must be the result of your own independent scientific judgement of the data or arguments or proofs and so on, and not be allowed to result from arguments based on reputation of others, nor upon such things as numerical strength of believers or disbelievers. When Einstein heard that a book was being brought out entitled "A Hundred Against Einstein", he merely said "One would be enough!" My own beads for Newtonian dynamics and Maxwell's equations are very near to 1, and for flying-saucers and the Loch Ness monster very near to 0. But these it must be emphasised are my own subjective beads, and it seems there exist people whose beads for UFOs are near to 1 or even at

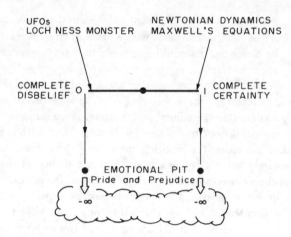

it and beyond, the consequences of which we proceed to discuss.

It seems to be a common defect of human minds that they tend to crave for complete certainty of belief or disbelief in anything. Not only is this undesirable scientifically, but it must be recognised that no such state is attainable in science. However successful and reliable a theory may be up to any point of time, further data may come along and show a need for adjustment of the theory, while at the other extreme, however little confidence one has in a hypothesis, new data may change the situation. We come now to the reason why one should never allow a bead ever to get right to 0 or 1: it is that, if one does so, the bead will fall into a deep potential-well associated with every facet of non-scientific or even anti-scientific emotion. In some cases the depth may tend to infinity, especially with advancing years, and no amount of data conflicting with the certain belief or disbelief will ever get the bead out of the well back onto the even tenor of the wire. Any attempt to bring about the uplifting of a bead so situated, by means of data or reason, can sometimes lead the owner of the bead to manifest further attitudes unworthy scientifically. In some cases it may be useless to discuss the hypothesis or theory to which the bead relates. On the other hand, if the bead is kept somewhere on the wire *between* 0 and 1 always, it can if necessary be moved quite readily in response to new data with the owner remaining calmly tranquil rather than undergoing an emotional upset. With such reaction to hypotheses and theories, one can get genuine scientific pleasure from adjusting one's beads to take account of new data and new arguments. From the small sample that my experience has limited me to, it seems regrettably to be the case that few even among scientists are always capable of keeping their beads on the wire, and much tact may be needed if one wishes to help to restore them to a rational level on the wire, if indeed in some cases it is possible at all. In Nazi Germany, it would have been dangerous indeed to have one's bead on the wire even near to 1 as an attitude to the theory that theirs was a super-race destined to rule the world; 99.9 per cent of the beads were deep down the well and only violent efforts proved sufficient to move some of them. So one of the things I would like to see scientists directed to do is always to keep their beads safely on the wire, in order that their minds may be receptive to new ideas and advances. In the words of one Chan, "Human mind like parachute: work best when open", and *open* means on the wire somewhere between 0 and 1.

When it comes to the question of how new ideas are to be arrived at, we meet up with the little-recognised fact that there is no such thing as "*the* scientific method": there is no formal procedure, no fixed set of rules, whereby new problems can be tackled or the correct interpretation of data in a new area rigidly attained, or whereby the necessary ideas to establish new principles can be reached by logical induction. However, the history of science shows that certain types of mind can see into a problem more deeply than others by some inner light that enables such minds to imagine what the solution might be, or to see if only gropingly at first what the required theory may be. It is not a matter of random guesswork, though it can be a kind of inspired guess or a series of guesses that reveals what may prove to be a valid theory for explanation of some phenomenon and for prediction in its realm.

Advancement of a subject may require first and foremost the intrusion into it of minds that intuitively perceive the phenomena in an original way and approach the problem from an untraditional point of view. With hindsight, it can be seen quite clearly just how impossible it was to solve the problem of the excess motion of the perihelion of Mercury by any hypothesis within the scope of then-existing astronomy. The imaginative conceiving of new ideas and the developing of them into a rational workable theory for new predictions ranks as mental activity at least the equal of those brought to bear in any of the realms of so-called arts, but where science is concerned it may go much further and lead on to yield a power over nature, not to mention intellectual joy and satisfaction, that no other art gives or possibly can give. It is this that makes science the noblest work of all and the greatest of the Arts. It is essentially this capability of knowledge of the future that is the keynote of real science, and the marvellous quality that gives science its uniquely powerful importance.

Before attempting construction of any theory (verbal or otherwise) to account for some phenomenon, it is of primary importance to establish that the phenomenon under discussion is actually occurring and not some wishful process swamped by the associated noise of measurement, for example. For what is to become of the theory if the phenomenon turns out later to have been purely imaginary and non-existent? Science journals today abound with elaborate theories of alleged phenomena quite inadequately established other than by often intemperate asseveration, and before the prime requirement of any theory has been found, namely an engine or cause of the phenomenon. The main evidence usually disseminated to support such "theory" often consists of no more than a sedulously conducted campaign of repetitious empty verbal propaganda assertively leading to the shallow conclusion that the theory is now "generally accepted", a claim in itself always a clear warning to regard the validity of the theory with reserve and to examine its basis (if any) calmly and independently paying no heed to the alleged numerical strength of its adherents. It is an essential part of scientific investigation to bring every detail of assumption, approximation, method, and all else to the surface, and have every component on the table, as it were, for examination and discussion: nothing should remain buried or left aside that any consideration suggests relevant until its importance or otherwise can be assessed. Equally so, in presenting scientific research publicly, it should be in such manner that a reader can recover for himself all the steps by which the results have been reached so that he can if necessary verify the conclusions for himself. In other words, none of the cards should remain face-down with assurances (or excuses) that this or that step is "all right" and can be taken for granted, because it has been put on a machine for instance, or otherwise remains inaccessible to verification. Yet many papers are deficient in this respect.

It is imperative to realise that the test of any new hypothesis or theory cannot be made by any *prior* supposedly aesthetic considerations or by moral judgement of their seeming merits or demerits in terms of existing theories. This can only be done posterior, *after* the consequences have been correctly worked out to a stage at which comparison with properly interpreted observations can be made. A theory is to be judged acceptable solely to the extent both that its results accord satisfactorily with the existing data and that future observations predicted on the basis of the theory duly come to pass with pleasing accuracy. When continuous variables are concerned, which is the most frequent case, perfect accuracy of prediction is never attainable any more than perfect accuracy of observation is possible; only if pure counting in whole numbers is entailed could this be possible, and even then of course it may not be achieved if the theory is imperfect. How little this scientific attitude to new ideas is adopted reveals itself by the sort of absurd comments they have been known to provoke, criticisms springing from beads well off the wire, such as, "The probability of the initial assumptions being true . . .", or the following unequalled gem: "The author seems unaware that the problem may be conditioned by some effect as yet unknown to science." These are just two samples from a great many that have come the way of this writer and sent out in all seriousness by chosen representatives of so-called learned societies, and if protest is made one may be informed, truthfully enough alas, that the opinions are from the most eminent referees in the field, not the most competent: the two are not always the same. It is axiomatic scientifically that no meaning attaches to such usage of the word "true": only when the initially assumed ideas have been followed through to comparison with data or to verified predictions can the stage arrive for their truth to be assessed. And just how an author can be expected to make allowance for "effects as yet unknown to science" requires an alchemy yet unknown to anyone.

Such absurdities would be avoided if the importance of keeping one's beads on the wire were sufficiently appreciated, and if it were also remembered that even though an idea should eventually turn out not to be true, in the sense that its predictions, properly evaluated of course, do not accord with observation, this alone does not constitute rigorous proof of the moral obliquity of the proposer. So reluctant are some scientists to bear such personal criticism that they shy away from the slightest risk of controversy, which new ideas frequently lead to, and devote themselves slavishly to the cloistered shelter of making routine measurements and observations, sometimes with no discernible objective. Some even make a virtue of the unimaginative collection of "facts", which they regard as the real work of science, and deride the searching for valid hypotheses as mere airy-fairy speculation. To dash a pail of water on the floor, and then set about the tedious measurement of the size and shape of all the splashes, with extensive tabulation of these, and published at great expense, would not represent in smallest degree a contribution to hydrodynamics, and obviously could never suggest the Stokes–Navier equations. Indeed, if anything, such misguided effort would far more likely prove an obstacle to any such important theoretical advance. Yet much modern work is in this sort of vein, and woe betide the theorist whose work takes no account of the resulting well-established "facts".

Any dedicated scientist will continually strive to imagine new ideas even though he recognises that most of them will probably not prove fruitful, but for every nine failures, one real success will be more than adequate reward for his efforts. But he will take care to test them out for himself by thinking them through in private, or by informal discussion with colleagues, or by observation or experiment, or when possible by formal analysis. Should these steps so reduce confidence in an idea to a point making further work upon it seem no longer worth while, then he must start over and try to imagine some new or modified idea in the hope that it may bring the data into order and suggest an appropriate theory. However, so great is the incentive to publish nowadays, that many regard it as good "scientific" practice to conduct this part of their education in public, mistakes and all. As a result, scientific literature becomes cluttered with inadequately ripe or even entirely erroneous material, which if subjected to open criticism is often, through faulty positioning of the relevant beads, quite invalidly defended, so great is the hurt to the pride of the authors and so unwilling are they to admit to the slightest error.

By way of conclusion and emphasis of the theme of this essay may be quoted what Poincaré had to say more than fifty years ago concerning an attempt to explain away the so-called ultra-violet catastrophe by means of a "theoretical" structure of matter more suggestive of poor-quality plumbing than the high-quality physical theory that eventually prevailed and revolutionised science: "It is obvious that by giving suitable dimensions to the communicating tubes connecting the reservoirs and giving suitable values to the leaks, this "theory" could account for any experimental results whatever. But this (type of descriptive story) is not the role of physical theories. They should not introduce as many (or more) arbitrary constants as there are phenomena to be explained. They should establish connections between different experimental facts, and *above all they should enable predictions to be made.*" It is to precepts of this stamp, distilled from the wisdom of great thinkers, that I would wish to see the attention of research workers perennially and forcefully directed, for only by adherence to them and in no other way will science be advanced.

J. A. Wheeler

J. A. Wheeler

Professor of Physics at the University of Texas, Austin, from September 1976 and Joseph Henry Professor of Physics Emeritus at Princeton University.

A past President of the American Physical Society, member of the U.S. National Academy of Sciences and recipient of the Enrico Fermi award and the National Medal of Science.

In nuclear physics his principal work has been concerned with the alpha-particle model, nuclear rotation, resonating group structure, introduction of the scattering matrix, the mechanism of fission, the design of plutonium production reactors and thermonuclear devices, and the collective model of the nucleus. More recently he has been occupied with gravitation physics, gravitational collapse, geons, neutron stars, black holes, gravitational waves, quantum fluctuations in geometry, and the super-space description of Einstein's general relativity.

C. M. Patton

Received his A.B. in Mathematics at Princeton University in 1972 and his Ph.D. at the State University of New York at Stony Brook in 1976; American Mathematical Society Postdoctoral Fellow, Institute for Advanced Study, Princeton, New Jersey, 1977-.

IS PHYSICS LEGISLATED BY COSMOGONY?

*... time and space are modes by which we think and not conditions in which
we live*

A. Einstein

A brief account of the Rutherford Laboratory Conference is followed by a review of the three
levels of gravitational collapse and of the crisis confronted by all of physics in the classically
predicted singularity at the end of time. One proposed way out considers that the deterministic
collapse predicted by classical theory is replaced in the real world of quantum physics by a
"probabilistic scattering in superspace" in which the Universe, momentarily extremely small, is
heavily "reprocessed" in all its features, before starting a new cycle of big bang, expansion,
constraction and collapse. This "reprocessing model" of cosmology asks no questions about
cosmogony, about how the Universe came into being. The contrasting view is also considered,
that the Universe comes into being at the big bang and fades away at collapse, and that all laws
are forced on physics by the requirement that the Universe must have a way to come into being.
The $- + + +$ signature of spacetime, and the quantum principle, are two examples of how
physics might be conceived to be thus legislated by cosmogony. Any hope of finding a basis in
the quantum dynamics of geometry for the reprocessing model would appear to be mistaken.
Five lines of evidence argue that geometry is as far from giving an understanding of space as
elasticity is from giving an understanding of a solid. They also suggest that the basic structure is
something deeper than geometry, that underlies both geometry and particles ("pregeometry").
For ultimately revealing this structure no perspective seems more promising than the view that
it must provide the Universe with a way to come into being.

1. QUANTUM RADIANCE

Some conferences are forgotten; but how can this conference be regarded as anything but
historic? Here for the first time Stephen Hawking showed us how to calculate from first
principles the quantum radiance of a black hole. For the first time we had before our eyes a
macroscopic and in principle observable effect born from the union of quantum mechanics and
general relativity. For the first time we had a proving-ground where we could hope to see in
action all the many different ways of doing the coupled wave mechanics of fields and geometry,
to learn what each has to give, as we learned in earlier days about quantum electrodynamics in
the proving ground of the Lamb–Retherford level shift of hydrogen. That level shift is small, as
the Bekenstein–Hawking temperature ($\sim 2 \times 10^{-7}$ deg K for a black hole of solar mass) is
small. Yet one decade sufficed to measure the one quantum effect with fantastic precision. Who
shall say that a century will not be enough for ingenious observers from the worlds of physics
and astrophysics to make at least an order-of-magnitude determination of this marvellous new
effect, the quantum radiance of a black hole? And if, as Hawking suggests is at least
conceivable, there are small black holes around, formed in the only way that anyone has been
able to imagine, in the big bang itself, then the violent quantum burnout of one of these objects
near the end of its life gives, as he points out, a second way to check on the predicted radiance.

TABLE 1. Black hole collapse, and the big bang and collapse of the Universe, as predicted by classical geometrodynamics, compared and contrasted with classically predicted collapse of the atom

System	Atom (1911)	Universe (1970s)
Dynamic entity	System of electrons	Geometry of space
Nature of classically predicted collapse	Electron headed toward point-center of attraction is driven in a finite time to infinite energy	Not only matter but space itself arrives in a finite proper time at a condition of infinite compaction
One rejected "way out"	Give up Coulomb law of force	Give up Einstein's field equation
Another proposal for a "cheap way out" that has to be rejected	"Accelerated charge need not radiate"	"Matter cannot be compressed beyond a certain density by any pressure, however high"
How this proposal violates principle of causality	Coulomb field of point-charge cannot readjust itself with infinite speed out to indefinitely great distances to sudden changes in velocity of charge	Speed of sound cannot exceed speed of light: pressure cannot exceed density of mass-energy
A major new consideration introduced by recognizing quantum principle as overarching organizing principle of physics	Uncertainty principle: binding too close to centre of attraction makes zero-point kinetic energy outbalance potential energy: consequent existence of a lowest quantum state; cannot radiate because no lower state available to drop to	"Participator" replaces the "observer" of classical physics. It is impossible in principle to separate what happens to any system, even the Universe, from what this participator does. This principle of Bohr's of the "wholeness" of nature (d'Espagnat: "non-separability") may be expected to come to the fore in a new and far deeper form

2. THE ORIGIN OF THE UNIVERSE AND THE CRISIS OF COLLAPSE

Hawking's work puts us on the road to seeing more clearly how quantum effects come into the dynamics of geometry. It brings us into closer confrontation than ever with the greatest question on the books of physics: How did the Universe come into being? And of what is it made?

Many great discoveries have been made in the past 500 years since the birth of Copernicus. None ranks higher for the light it shed on existence than the discovery by Darwin and his successors of how present life forms came into being. No discovery penetrated mysteries more widely agreed to be forever beyond the power of the mind of man to fathom. No achievement gives more hope that the next 500 years hold in store for us a still greater discovery, how the Universe itself came into being.

If there is no hope of progress towards a discovery without a paradox, we can rejoice in the paradox of gravitational collapse and the associated paradox of the big bang. Let a computing machine calculate onward instant by instant towards the critical moment, and let it make use of Einstein's standard 1915 geometrodynamics. Then a point comes where it cannot go on. Smoke, figuratively speaking, rises from the machine. Physics stops. Yet physics has always meant that which goes on its eternal way despite all the surface changes in appearances. Physics stops; but physics goes on: here is the paradox.

Physics grappled once before with a comparable paradox. In the 1910s Ernest Rutherford had shown that matter is made of highly localized positive and negative charges. Then matter must collapse in a time of the order of 10^{-17} sec. But matter does not collapse. This paradox of collapse, not the orbit of the electron in the hydrogen atom, was the overriding concern of Niels Bohr month after month. Many proposed finding a way out by giving up the Coulomb law of force at small distances, or giving up the laws of electromagnetic radiation, or both. Bohr, in contrast, held fast to both. At the same time he recognized the importance of a third law, Planck's radiation law, that seemed at first sight to have to do exclusively with quite another domain of physics. Only in that way did he find the solution of the apparent paradox. We are equally prepared today to believe that a deeper understanding of the quantum principle will someday help us resolve the paradox of gravitational collapse (Table 1).

3. GEOMETRY: PRIMORDIAL OR DERIVATIVE?

Tied to the paradox of big bang and collapse is the question, what is the *substance* out of which the Universe is made? Great attraction long attached to W. K. Clifford's 1870 proposal, and Einstein's perennial vision, that space geometry is the magic building material out of which particles and everything else are made. That vision led one to recognize perhaps earlier than might otherwise have been the case how rich are the consequences of Einstein's standard geometrodynamics, including not least constructive properties sometimes epitomized in the shorthand phrases "mass without mass" (geons), "charge without charge" (charge as electric lines of force trapped in the topology of a multiply connected space), and "spin without spin" (distinction between, and separate probability amplitudes for, the 2^n distinct geometrical structures that arise out of one and the same 3-geometry, endowed with n "handles" or "wormholes", according as one or another topologically distinct continuous field of triads is laid down on that 3-geometry). However, in the end the explorations of the Clifford–Einstein space theory of matter have taught us the deficiencies of "geometry as a building material", and Andrei Sakharov has reinforced the lesson. No one sees any longer how to defend the view that "geometry was created on 'Day One' of creation, and quantized on Day Two". More reasonable today would appear the contrary view, that "the advent of the quantum principle marked Day One, and out of the quantum principle geometry and particles were both somehow built on Day Two".

Glass comes out of the rolling mill looking like a beautifully transparent and homogeneous elastic substance. Yet we know that elasticity is not the correct description of reality at the microscopic level.

Riemannian geometry likewise provides a beautiful vision of reality; but it will be as useful as anything we can do to see in what ways geometry is inadequate to serve as primordial building material.

Then, at the end of this account, it will be appropriate to turn to the quantum principle as primordial, or as a clue to what is primordial. There our objective must of necessity be, not the right answers, but a start at the more difficult task of asking the right questions.

4. "QUANTUM GRAVITY"

What is sometimes called "quantum gravity" has a different and more immediate objective. Take as given Einstein's Riemannian space and his standard classical geometrodynamic law; and investigate the quantum mechanics of this continuous field, as one does for any other continuous field.

However workable this procedure of "quantization" is in practice for some fields and most discrete systems, we know that in principle it is an inversion of reality. The world at bottom is a quantum world; and any system is ineradicably a quantum system. From that quantum system the so-called "classical system" is only obtained in the limit of large quantum numbers. It is an accident peculiar to a sufficiently simple system that for it "the circle closes exactly"; for example, (1) quantum harmonic oscillator to (2) its limiting behaviour for large quantum numbers to (3) the "classical harmonic oscillator" to (4) the so-called "process of quantization" back to (5) the original quantum system. In contrast to this mathematics, nature does not "quantize"; it is already quantum. Quantization is a pencil-and-paper activity of theoretical physicists. This circumstance gives us always some comfort when a quantum field theory turns out to be "unrenormalizable", because we know that for nature the word "renormalization" does not even exist; nature manages to operate without divergences!

However hopeful these general considerations may be for the future they are of scant help in getting on with the immediate tasks, and tracing out the immediate consequences, of quantum gravity. No one today knows how to get quantum theory as quantum theory without having at the start the mathematical guidance of what we call a "classical" theory. No one has found a way, and nobody sees a need to find a way, to talk about quantum gravity that does not constantly make use of one or another of the concepts of canonical quantization or covariant quantization, from canonical field momentum to radiative degrees of freedom and initial value data, and from "background plus fluctuations", and "closed loops plus tree diagrams", to superspace. Abdus Salam reminded us of the important similarities and also the important differences between quantum geometrodynamics, and quantum field theory as developed in the context of Minkowski spacetime. His interesting and far-reaching survey mentioned among other points eight important "inventions" that have been made in elementary particle theory in recent years. It makes him and us happy that in quantum gravity one has as "Lehrbeispiel" a field theory that carries a far lighter burden of arbitrary elements.

For the mathematical framework in which a field theory expresses itself, one has been familiar for a long time with such alternatives as a direct description in spacetime, a description in terms of Fourier amplitudes, and a description in terms of the scattering matrix; but the latter two alternatives give a full payoff only in the context of flat spacetime. Roger Penrose and George Sparling showed in twistors quite another formalism for doing field theory, still in Minkowski geometry, but one that conceivably will let itself be extended or generalized to deal with curved space and its dynamics — if so to our great enlightenment.

The mathematical methods of quantum gravity have many beauties. Hawkings' arguments for and analysis of black hole radiance are history-making. Not only on these two counts does special interest attach to our subject today, but also because new X-ray and other astrophysical observations make it seem quite possible that the compact X-ray source Cygnus X-1 is the first identified black hole.

5. QUANTUM FLUCTUATIONS AND THE THIRD LEVEL OF GRAVITATIONAL COLLAPSE

The black hole, as "experimental model" for gravitational collapse, brings us back full-circle to the paradox that continually confronts us, and all science, the paradox of big bang and gravitational collapse of the Universe itself. The existence of these two levels of collapse reminds us, however, that theory gives us also what is in effect a third level of collapse, small-scale quantum fluctuations in the geometry of space taking place and being undone, all the time and everywhere.

Among all the great developments in physics since World War II, there has been no more impressive advance in theory than the analysis of the fluctuations that take place all the time and everywhere in the electromagnetic field. There has been no more brilliant triumph of experimental physics than the precision measurement of the effect of these fluctuations on the energy levels of the hydrogen atom. There has been no more instructive accord between observation and theory.

These developments tell us unmistakably that the electron in its travels in a hydrogenic atom is subject not only to the field Ze/r^2 of the nucleus, but also to a fluctuation field that has nothing directly to do with the atom, being a property of all space. In a region of observation of dimension L the calculated fluctuation field is of the order,

$$\Delta\epsilon \sim (\hbar c)^{\frac{1}{2}}/L^2. \tag{1}$$

This field causes a displacement Δx of the electron from the orbit that it would normally follow. This displacement puts the electron at a slightly different place in the known potential $V(x,y,z)$ in which it moves. In consequence the electronic energy level undergoes a shift,

$$\Delta E \simeq [(\Delta x)^2/2] < \nabla^2 V > \text{average}. \tag{2}$$

Measuring the main part of the Lamb–Retherford shift, one determines the left-hand side of this equation. One knows the second term on the right-hand side. Thus one finds the mean squared displacement of the electron caused by the fluctuation field and confirms the predicted magnitude of the fluctuation field itself.

The considerations of principle that give one in electrodynamics the fluctuation formula (1) tell one that in geometrodynamics, in a probe region of extension L, the quantum fluctuations in the normal metric coefficients $-1, 1, 1, 1$ are of the order,

$$\Delta g \sim L^*/L. \tag{3}$$

Here

$$L^* = (\hbar G/c^3)^{\frac{1}{2}} = 1.6 \times 10^{-33} \text{ cm} \tag{4}$$

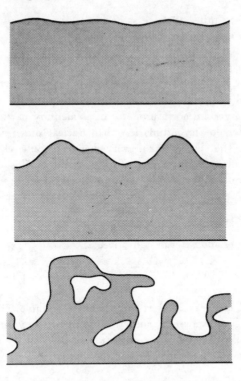

Fig. 1. Schematic representation of typical fluctuations in the geometry of spaces as picked up at three levels of observation, ranging from a large scale of observation, L, at the top, to a scale of observation comparable to the Planck length at the bottom.

is the Planck length. These fluctuations are negligible at the scale of length, L, of atoms, nuclei, and elementary particles, as the wave-induced fluctuations in the level of the ocean appear negligible to an aviator flying 10 km above it. As he comes closer, or as L diminishes, the fluctuations become more impressive (Fig. 1). Finally, when the region of analysis is of the order of the Planck length itself, the predicted fluctuations are of the order $\delta g \sim 1$. When the changes in geometry are so great, no one sees how to rule out the possibility that there will also be changes in connectivity (bottom frame in Fig. 1), with "handles" or "wormholes" in the geometry all the time and everywhere forming and disappearing, forming and disappearing ("foam-like structure of space").

One who had never heard of electricity, looking for evidence of this multiple connectivity of space, would *predict* electricity as consequence of it. Thereupon *finding* electricity in nature, he would take this discovery as evidence that space really is multiply connected in the small. Nothing prevents our rising above the accidents of history to take the same position.

The concept that electricity is lines of force trapped in the topology of a multiply connected space was put forward by Hermann Weyl in 1924, before one even knew about wave mechanics, let alone quantum fluctuations in geometry. Today quantum theory tells us that electric lines of force of fluctuation origin will thread through the typical wormhole of dimension L, carrying a flux of order

$$\int E \cdot dS \sim [(\hbar c)^{1/2}/L^2] \cdot L^2 \sim (\hbar c)^{1/2}. \tag{5}$$

Accordingly we are led to think of space as having a kind of fluctuating foam-like structure, with everywhere positive and negative charges of order

$$q \sim (\hbar c)^{\frac{1}{2}} \sim 10e \tag{6}$$

continually being created and annihilated.

These fluctuation charges are not a property of elementary particles. The relevant scale of distances is twenty orders of magnitude less than nuclear dimensions. The charges are not quantized in magnitude. The charges occur everywhere, not only where there is a particle. The charges are exclusively electric in character. In evidence of this conclusion, note that the primary quantity in the quantum-electrodynamics analysis is the potential A. The equation

$$B = \nabla \times A,$$

or, in the language of forms and exterior derivative,

$$B = dA, \tag{7}$$

excludes the trapping of magnetic lines of force in any wormhole. Thus, drape a closed surface S with the topology of a 2-sphere around the mouth of a wormhole, and note that the one-dimensional boundary, ∂S, of S is automatically zero. Then the magnetic pole strength, p, associated with this wormhole mouth is given by the flux integral

$$4\pi p = \int_S B = \int_S dA = \int_{\partial S} A \equiv 0. \tag{8}$$

Thus the same wormhole concept that gives electricity rules out magnetic poles.

Why fluctuation charges should assemble into elementary particles, and why the net charge of each assembly should be quantized, are, like the question why should there be any elementary particles at all, beyond our power to answer today.

In times past the assumptions were tacitly made (1) that the concept of "space" makes sense at small distances and (2) that the topology, or connectivity, of space in the small is Euclidean. These assumptions left no escape from the picture of charge as a mystic magic electric jelly, or the equally unsatisfactory picture of charge as associated with a place where Maxwell's equations break down.

No one has ever been able to explain why there should be such an entity as electricity, and why there should not be such an entity as a magnetic pole, except by giving up the assumption that connectivity in the small is Euclidean, and accepting the concept of lines of force trapped in the topology of space. Nothing gives one more reason to take seriously the prediction of something like gravitational collapse going on all the time and everywhere, at small distances, and continually being done and undone, than the existence of electricity.

The value of the scale dimension L of the typical wormhole cancels out in expression (5) for the fluctuation charge itself. Moreover, the magnitude of the fluctuation charge has nothing whatsoever directly to do with the magnitude e of what on these views is an assembly of such charges. Therefore nothing can be learned from the *magnitude* of the quantum of charge about the distance down to which the concept of geometry makes sense. However, from the *existence* of charge it would seem to follow on this picture, either that the fluctuations in geometry are great enough to bring about changes in topology ($L \sim L^*$ in eqn. (3)) or that the very concept of geometry fails in such a way at small distances (distances L^* or larger) as to bring about the equivalent of a change in connectivity.

The view that large fluctuations go on at small distances puts physics in a new perspective. The density of mass-energy associated with a particle (one or a few orders of magnitude larger than the density of nuclear matter, 2×10^{14} g/cm^3) is as unimportant compared to the calculated effective density of mass-energy of vacuum fluctuations down to the Planck scale of lengths,

$$\rho \sim \frac{[(\hbar c/L^*)/c^2]}{L^{*3}} \sim \frac{M^*}{L^{*3}} \equiv \frac{2.2 \times 10^{-5} \text{ g}}{(1.6 \times 10^{-33} \text{ cm})^3} \sim 10^{94} \text{ g/cm}^3, \tag{9}$$

as the density of a cloud, $\sim 10^{-6}$ g/cm^3, is unimportant compared to the density of the sky, $\sim 10^{-3}$ g/cm^3. The track of the particle looks impressive on its passage through a Wilson chamber. The white cloud, too, looks impressive in the transparent sky. However, the proper starting point in dealing with physics in the one case is the sky, not the cloud; and we are free to believe that the proper starting point in the other case is the physics of the vacuum, not the physics of the particle. To adopt this perspective does not yield any sudden illumination about either particles or the vacuum, but does at least suggest that no theory of particles that deals only with particles will ever explain particles.

6. RELATION BETWEEN THE THREE LEVELS OF COLLAPSE

Relegate to Appendix A† a few notes about the mathematical machinery for describing quantum fluctuations in the geometry and the quantum state of the geometry, and turn directly to the connections between the three levels of gravitational collapse that we have just reviewed. For definiteness adopt Einstein's view that the Universe is closed. Then we expect a spacelike singularity in the geometry at collapse similar to the spacelike singularity predicted at the big bang. Thus a black hole, once formed, cannot endure for ever. The singularity associated with it ultimately has to amalgamate with the cosmological singularity, as the icicle hanging from the roof of an ice cave, traced upward from its tip, has to join on to that roof. It does not change this conclusion to have several black holes join together into fewer and larger black holes, either early on, or as part of the final collapse itself.

The irregularities in the overall 3-geometry occasioned by black holes and other compact objects in no way prevent the systematic following out of the steady dynamics of contraction as depicted in terms of the "Kuchař–York extrinsic time",‡ τ. For any finite value of τ, however large, the spacelike hypersurface on which the 3-geometry is being studied, however close it may be to the final spacelike cosmological singularity, has still so perfectly accommodated itself to "the icicles hanging from the ceiling" that it does not quite touch the tip of any one of them. Moreover, the proper volume of this hypersurface of constant τ, as

† For reasons of brevity Appendix A has not been included in this work.

‡ This "extrinsic" time is defined in Appendix A, equation (A25). For a simple Friedmann universe of radius $a = a(t)$, at late times, one has

$$\tau = -(4/a)(da/dt) = [(8/3) \text{ or } 2]/(t_{\text{final}} - t), \tag{10}$$

according as the model is matter-dominated or radiation-dominated.

calculated classically, decreases indefinitely as τ increases indefinitely in the final stages of collapse. This circumstance links collapse of the Universe, and collapse to a black hole, not only with each other, but also with the third level of collapse, the collapse predicted to be all the time taking place and being undone at small distances. The classical prediction of a zero volume at infinite τ therefore surely loses force for real physics when dimensions as calculated classically have fallen to the order of magnitude of the Planck length, if not before.

7. PROBABILISTIC SCATTERING IN SUPERSPACE AND THE "REPROCESSING MODEL" OF COSMOLOGY

Why not take a model universe, for simplicity even one that derives its entire content of effective mass-energy from gravitational waves and source-free electromagnetic fields, and let the quantum dynamics of this system systematically crank ahead through the phase of collapse to whatever happens afterwards? The electron travelling towards a point centre of positive charge arrives in a finite time at a condition of infinite kinetic energy, according to classical theory, just as the Universe arrives in a finite proper time at a condition of infinite compaction. But, for the electron, quantum theory replaces deterministic catastrophe by probabilistic scattering in (x,y,z)-space. Why then for the Universe should not quantum theory replace deterministic catastrophe by probabilistic scattering in superspace? Even without the actual quantum geometrodynamic calculation, which is too difficult — and too difficult to define — for today's power of analysis, can one not conclude that any given cycle of expansion and contraction is followed, not by a unique new cycle, but a probability distribution of cycles? According to this expectation, in one such cycle the Universe attains one maximum volume and lives for one length of time; in another cycle, another volume and another time; and so on. In a few such cycles life and consciousness are possible; in most others, not; no matter; on this view the machinery of the Universe has nothing to do with man. In brief, this picture considers the laws of physics to be valid far beyond the scale of time of a single cycle of the Universe, and envisages the Universe to be "reprocessed" each time it passes from one cycle to the next (model of "reprocessing within the framework of forever frozen physical laws").

8. CONTRASTING MODEL OF "PHYSICS AS LEGISLATED BY COSMOGONY"

A universe built of geometry (and fields), with this geometry ruled by Einstein's field equation (plus quantum mechanics), is central to this "reprocessing model" of cosmology. However, an examination in §10, from five points of view, will argue that (1) "geometry" is as far from giving an understanding of space as "elasticity" is from giving an understanding of a solid; and (2) a geometrodynamic calculation extended through the regime of collapse, if extend it one ever can, will give results as misleading as those from a calculation via elasticity theory on the fracture of iron. It is difficult to escape the conclusion that "geometry" must be replaced by a more fundamental concept, as "elasticity" is replaced by a collection of electrons and nuclei and Schrödinger's equation. Whatever the deeper structure is that lies beneath particles and geometry, call it "pregeometry" for ease of reference, it must be decisive for what goes on in the extreme phases of big bang and collapse. But is it really imaginable that this deeper structure of physics should govern how the Universe came into being? Is it not more reasonable to believe the converse, that the requirement that the universe should come into being governs the structure of physics? If so, the problem of "pregeometry" is not separate from the problem of cosmogony; the two are aspects of one and the same problem.

As an illustration of what it might mean to speak of "physics as legislated by cosmogony", look at the difference between a spacetime with metric in the local tangent space of the form

$$- dt^2 + dx^2 + dy^2 + dz^2 \tag{11a}$$

and one of the form

$$+ dt^2 + dx^2 + dy^2 + dz^2. \tag{11b}$$

Even in the simplest case of an ideal spherical model universe the resulting alteration is drastic in the history of the radius, a, as a function of time: from

$$(da/dt)^2 - (a_{max}/a) = - 1, \tag{12a}$$

with solution

$$a = (a_{max}/2) \, (1 - \cos \eta),$$

$$t = (a_{max}/2) \, (\eta - \sin \eta), \tag{13a}$$

to

$$(da/dt)^2 + (a_{min}/a) = + 1, \tag{12b}$$

with solution

$$a = (a_{min}/2) \, (1 + \cosh \eta),$$

$$t = (a_{min}/2) \, (\eta + \sinh \eta), \tag{13b}$$

where in both cases, for simplicity, pressure is treated as negligible, as is appropriate for a mixture of stars and dust. The scale of the Universe, far from being defined by a maximum radius, is defined by a minimum radius (Fig. 2). No longer does the history show big bang or collapse. With no big bang there is no singularity, no "umbilicus", and no evident way for such a universe ever to come into being.

Shall we conclude that the only cosmology worse than a universe with a singularity is a universe without a singularity, because then it lacks the power to come alive? Is this why nature rules out a positive definite metric for spacetime? If so, this is an example of "the structure of physics as forced by the requirement for the coming into being of the universe".

The quantum principle may conceivably some day provide a second example of what it might mean to think of "physics as legislated by cosmogony". Quantum mechanics does not

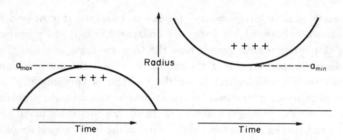

Fig. 2. Radius of an ideal spherical model universe as a function of time: left, for the $-+++$ signature of the real world; right, for $++++$ signature.

supply the Hamiltonian; it asks for the Hamiltonian; but beyond the rules of quantum mechanics for calculating answers from a Hamiltonian stands the quantum principle. It tells what question it makes sense for the observer to ask. It promotes observer to participator (Fig. 3). It joins participator with system in a "wholeness" (Niels Bohr) or "non-separability" (Bernard d'Espagnat) quite foreign to classical physics. It demolishes the view we once had that the Universe sits safely "out there", that we can observe what goes on in it from behind a foot-thick slab of plate glass without ourselves being involved in what goes on. We have learned that to observe even so miniscule an object as an electron we have to shatter that slab of glass. We have to reach out and insert a measuring device (Fig. 3). We can put in a device to measure position or we can insert a device to measure momentum. But the installation of the one prevents the insertion of the other. We ourselves have to decide which it is that we will do. Whichever it is, it has an unpredictable effect on the future of that electron. To that degree the future of the Universe is changed. We changed it. We have to cross out that old word "observer" and replace it by the new word "participator". In some strange sense the quantum principle tells us that we are dealing with a participatory universe.

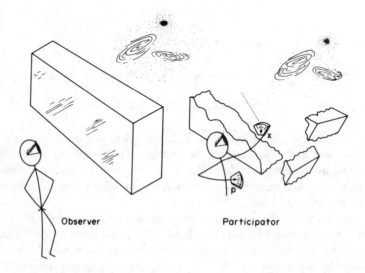

Fig. 3. The quantum principle throws out the old concept of "observer" and replaces it with the new concept of "participator". It demolishes the concept that the Universe sits "out there".

9. SELF-REFERENCE COSMOGONY

Is the necessity of this principle in the construction of the world beyond hope of explanation? Or does it, along with $-+++$ signature, originate in the requirement for the coming into being of the Universe? If so, the strange and inescapable role of observer-participator in physics cannot easily be imagined to come from anything but a strange and inescapable role for the observer-participator in cosmogony itself. The idea is very old (Parmenides of Elia, $\sim$ 500 B.C.; George Berkeley, $\sim$ 1710) that the "observer" gives the world the power to come into being, through the very act of giving meaning to that world; in brief, "No consciousness; no communicating community to establish meaning? Then no world!" On this view the Universe is to be compared to a circuit self-excited in this sense, that the Universe gives birth to consciousness, and consciousness gives meaning to the Universe (Fig. 4).

"In giving meaning to the Universe, the observer gives meaning to himself, as part of that Universe." With such a concept goes the endless series of receding reflections that one sees in a pair of facing mirrors. In this sense one is dealing with "self-reference cosmogony".

No test of such thinking would be more decisive than a derivation of the quantum principle from this "self-reference cosmogony", as one long ago derived the formula for the energy of a moving electron from "relativity". For developing a mathematical description of this new kind of "meaning circuit" one has as guides: (1) what one already knows from quantum mechanics; (2) what one has learned from the central role of self-reference in the revolutionary discoveries of recent decades in mathematical logic — the only branch of mathematics that has the power "to think about itself"; and (3) experience in analysing physical situations that transcend time and bring past, present and future together under one roof.

As contrasted to the cycle-after-cycle "reprocessing model" of cosmology — which is indifferent to how the world came into being — self-reference cosmogony has these features: (1) one cycle only; (2) the laws and constants and initial conditions of physics frozen in at the big bang that brings the cycle and dissolved away in the final extremity of collapse;† and (3) a guiding principle of "wiring together", past, present, and future that does not even let the Universe come into being unless and until the blind accidents of evolution† are guaranteed to produce, for some non-zero stretch of time in its history-to-be, the consciousness, and consciousness of consciousness, and communicating community, that will give *meaning* to that universe from start to finish.

Of all evidence that might distinguish between the two models, one the reprocessing of a forever existing universe, the other the coming-into-being and fade-out of a one-cycle-only universe, nothing attracts inquiry more immediately than geometry. Is space really a 3-manifold? And is its quantum dynamics really ruled by Einstein's field equation? "Yes" is the working conception of the reprocessing model. "No" has to be the answer of self-reference cosmogony. True mathematical 3-geometry "is simply there"; it has no way to "come into

---

† On this view of "big bang as coming into being, collapse as dissolution" there is nothing "before" the moment of commencement of the Universe, and nothing "after" the moment of collapse, not least because already in the 3-geometry-superspace formulation of quantum geometrodynamics all meaning is denied to the terms "before" and "after", and even to "time", in the analysis of (1) what goes on at sufficiently small distances and therefore (2) what happens to the Universe itself when *it* is sufficiently small.

† Chance mutation, yes; Darwinian evolution, yes; yes, the general is free to move his troops by throwing dice if he chooses; but he is shot if he loses the battle. Deprived of all meaning, stripped of any possibility to exist, is any would-be universe where Darwinian evolution brings forth no community of evidence-sharing participators, according to the view of "self-reference cosmogony" under examination here.

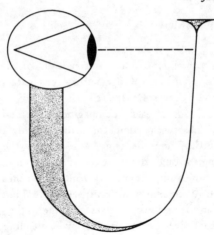

Fig. 4. Symbolic representation of the Universe as a self-excited system brought into being by "self-reference". The Universe gives birth to communicating participators. Communicating participators give meaning to the Universe.

being". If the world contains any such ingredient, the idea is contradicted from the start that all physics, all structure, and all law arise from the requirement that the Universe must have a way to come into being. Thus the relevance is clear of a closer look at the role of geometry in physics.

To suppose that the attention of all investigators is now limited, or ever will be limited, to the two cited concepts of the Universe, would be to underestimate the exploratory character of the community of science, that makes it so effective in uncovering new evidence, winning unexpected insights, and achieving final consensus. More ideas and more cosmologies will surely receive attention, not least because our understanding of the tie between cosmogony and the basic structure of physics ("structural cosmogony") is today so clearly in its medieval infancy. The road ahead can hardly help being strewn with many a mistake. The main point is to get those mistakes made and recognized as fast as possible! It may help in this work to reassess the concept of geometry, regardless of its bearing on reprocessing cosmology or self-reference cosmogony.

10. REASSESSMENT OF "GEOMETRY"

Five considerations suggest that 3-geometry does not give a correct account of physics and ought to be replaced by a more basic structural concept ("pregeometry"): (1) Mathematical space doesn't tear, but physical space must tear. (2) Mathematical space has only geometrical properties ("surface geology") but physical space is full of virtual pairs of all kinds of particles ("underground strata"). (3) It is difficult to avoid the impression that every law of physics is "mutable" under conditions sufficiently extreme, and therefore that geometry itself — a part of physics — must also be mutable. (4) Einstein's law for the dynamics of geometry follows from nothing deeper than the "group" of deformations of a spacelike hypersurface (Hojman, Kuchař and Teitelboim) and therefore tells nothing more about the underlying structure than crystallography tells about the structure of an atom. (5) Gravitation, as the "metric elasticity of space" (Andrei Sakharov), is as far removed from the deeper physics of space as elasticity itself is from the deeper physics of a solid.

"Doesn't Tear; Must Tear"

Nothing speaks more strongly than the existence of electricity, and the non-existence of magnetic poles, for small-scale fluctuations in the connectivity as well as in the geometry of space. What are the consequences of such changes in connectivity? When a handle thins and breaks, two points part company that were once immediate neighbours. But in quantum physics no change is sudden. There must be some residual connection between these two points. Moreover, there is nothing special about these two points. If there is a physical tie between them, there must also be a physical tie between every point and every other point that is very foreign to differential topology. The concept of nearest neighbor would seem no longer to make any sense; and with it gone, even the foundation for the idea of dimensionality disappears. Farewell geometry!

"Surface versus Stratigraphic Geology"

When sufficient electromagnetic energy is imploded into a sufficiently small volume of space, pairs of particles emerge. The "stratigraphic structure" built into the vacuum, and thus revealed, cannot be disregarded merely because it does not seem to lend itself to description in geometrical language, however different it is in kind from the Riemannian curvature ("surface geology") that describes gravitation. Moreover, if gravitation-as-curvature is universe-wide, so is this stratigraphy, to our astonishment. How did a quasar at red-shift $z = 3$, so separated from us in place and time, "know enough" to show by its spectrum the same atomic properties as an atom in the here and now? The elementary particles in the source at the time of emission had been around only $\sim 1 \times 10^9$ yr since the big bang, whereas they would have had to reach an age $\sim 2 \times 10^9$ yr to get the first information from us about particle properties here. Zel'dovich gives a beautiful description of pair production, showing how the particles come into evidence at a spacelike separation of the order of magnitude of the Compton wavelength. No way is evident of reconciling with the principle of causality the identity of stratigraphy between points with spacelike separations of 10^9 ℓyr and 10^{10} ℓyr except to say that they also once had separations of the order of a Compton wavelength. Whatever this consideration does to explain how mass-spectral information imprinted on space at different points can be identical, it does nothing to explain how the imprinting itself is possible. That is a property of empty space quite irreconcilable, so far as we can see, with any known concept of geometry whatsoever. Farewell geometry.

Mutability

Physics can be viewed as a staircase. Each tread marks a new law. Each riser symbolizes the achievement of conditions sufficiently extreme to overpower that law. Density seemed a conserved quantity until one discovered that sufficiently great pressures alter density. Valence won ever wider fields of application until one found that temperatures can be raised so high that the concept of valence loses its usefulness. The fixity of atomic properties is transcended in thermonuclear reactions. The principle of the conservation of lepton numbers and baryon numbers, discovered to be central to elementary particle physics, is transcended in the *classical* physics of black holes. The black hole obeys in its transformations the laws of conservation of mass-energy, electric charge, and angular momentum. However, none of these would-be conserved quantities has any meaning when one turns from a finite system to a closed universe. Finally, with the classically predicted collapse of the Universe, space and time themselves are transcended as categories, and the framework would seem to collapse for everything one has ever called a law, not least the concept of geometry itself. Farewell geometry.

Groups Conceal Structure

Pile up a 50-m-high pyramid of wrecked automobiles and steel bed springs, attach a spring to the top, drive it with slowly changing period, and measure the characteristic frequencies. Each mode of vibration is orthogonal to all the others, but not one clue does this fact, or all the details of the frequency spectrum, provide as to the constitution of the vibrating system. Nor does the cubic symmetry of salt reveal anything about the constitution of the sodium atom. Turning from the sodium atom to other selected atoms, and analysing the pattern of the lowest score or so of energy levels — H. P. Dürr has shown — one would conclude that the system in question has SU(3) or higher symmetry; but nothing could be more ridiculous than to conclude from this circumstance that the outer structure of the atom is made up from quarks or from hadrons.

More generally, much that appears to be "structure" turns out on more detailed analysis to be mere consequence of the principle that "a symmetry is spontaneously broken when for a physical problem invariant under a group G there exist solutions (which can be grouped into orbits of G) which are only invariant under a strict subgroup of G"; hence features of physics like the Jahn–Teller effect, loss by crystals of certain symmetries, spontaneous magnetization, and the Cabibbo angle for the weak current. Groups hide structure! Similarly in general relativity.

All the laws of gravitation follow from the Einstein–Hamilton–Jacobi equation, and this equation follows from a beautiful argument of group theory (invariance with respect to the "group" of deformations of a spacelike hypersurface; in evolving dynamically on σ from σ' to σ'', the system must end up in the same state on σ'' whether σ was pushed ahead faster, first "on the left", then "on the right", or faster first on the right, then on the left). The very fact that group theory is the heart of this derivation of the dynamics of geometry, more than concealing from view a structure deeper than "ideal mathematical geometry", would seem a prime sign that there must be such an underlying structure. Farewell geometry.

Gravitation as the "Metric Elasticity of Space"

Is geometry the magic building material out of which everything is made? To try to build particles out of geometry is as mistaken, Sakharov in effect argues, as to try to build atoms out of elasticity; the plan of construction goes the other way around. The two elastic constants of a homogeneous isotropic material not only summarize but also hide from view the second derivatives of the hundred atomic and molecular bonds that are the source of that elasticity. Sakharov proposes to look at gravitation in similar terms. Space, in his view, if we may coin an analogy, is like an empty sausage skin. It puts up no resistance to being bent until it is pumped full of sausage meat. This sausage meat is the zeropoint energy of particles and fields. When space is curved, correction terms arise in the renormalized invariant Lagrangian density for each field, proportional to the four-dimensional Riemann scalar curvature invariant,

$$\delta L_{\text{one field}} \sim \hbar c^{(4)} R \int_0^{k} {}^{\text{cut off}} k \, dk; \qquad (14)$$

and with the cut-off wave number taken to be of the order of the reciprocal of the Planck length (to give the right answer!) this agrees in form and general magnitude with the Lagrangian for the gravitational field itself. The constant of gravitation, on this view, measures the "metric elasticity" that space acquires from its content of particles and fields.

Though a step forward, it was not a simplification to replace two elastic constants of a solid by the potential energy curves of a hundred molecular bonds. The real simplification came when one recognized the solid as nothing but a system of charged masses moving in accordance with the laws of quantum mechanics. The thousand remarkable features of chemistry and chemical physics found their explanation, not in a thousand specially tailored laws, but one simple picture. Neither can we believe that Sakharov has gone the whole road when he explains the resistance of geometry to bending as a consequence of the complicated physics of dozens of varieties of particles and fields. Neither particles explained in terms of geometry, nor geometrodynamics explained in terms of particles, gives the truly simple concept, we have to believe, but both entities expressed in terms of a structure deeper than either, call it pregeometry or call it what one will. Farewell geometry.

In conclusion, it would appear from these five considerations of tearing, stratigraphy, mutability, group concealment, and metric elasticity that the concept of "ideal mathematical geometry" as applied in physics is too finalistic to be final and must give way to a deeper concept of structure. Towards the finding of this "pregeometry" no guiding principle would seem more powerful than the requirement that it should provide the Universe with a way to come into being. It is difficult to believe that we can uncover this pregeometry except as we come to understand at the same time the necessity of the quantum principle, with its "observer–participator", in the construction of the world. Not "machinery", but a guiding principle, is what we seek. It would seem a great mistake to expect a "cheap solution" of the crisis of collapse. Does this mean that we must decipher almost everything before we will be able to understand anything? Certainly of problems of less than cosmic magnitude in gravitation physics and quantum gravity there are more, and more of greater interest, outstanding today, than at any time we can name. They range from renormalization to the mathematics of topology change, and from the hydrodynamics and gravitational radiation associated with the collapse of a rotating star, to the Hawking radiation and the polarization of the vacuum around a black hole. As for the problem of "structural cosmogony", the "great" problem, we are reminded that every great crisis in science has created its own hard conditions — and opportunities — for making progress. In looking for a satisfactory model to bring into harmony the rich evidence that nature lays before us, perhaps we can derive some comfort from the words of the engine inventor, John Kris, "Start her up and see why she don't go".

Acknowledgement

This article originally appeared in *Quantum Gravity*, edited by C. J. Isham, R. Penrose and D. W. Sciama, Clarendon Press, Oxford, 1975. It included two appendices, bibliography, references and acknowledgements, all of which are omitted here in the interest of brevity. —Ed.

I. W. Roxburgh

Professor of Applied Mathematics at Queen Mary College, University of London, and Chairman of the London University Astronomy Committee.

Member of the Royal Astronomical Society and the International Astronomical Union, Member of the Council of the Royal Astronomical Society 1968-71.

Research interests include the History and Philosophy of Science, Cosmology, Evolution of Stars, Solar Physics and Gravitational Theory.

THE COSMICAL MYSTERY— THE RELATIONSHIP BETWEEN MICROPHYSICS AND COSMOLOGY

Is the Universe of which we are a part unique, or is it just one out of a whole family of possible Universes? Is the world as it is because there is no other way for it to be, or because a Creator arbitrarily chose to create it this way rather than another? Can physics help to answer these questions?

The laws of physics so far uncovered by man contain arbitrary dimensionless numbers whose values determine the properties of the Universe; change these values and the properties of the Universe change, so at present physics suggests an ensemble of possible Universes. But are these numbers arbitrary or is it that we have yet to discover the interrelatedness of nature and how this uniquely determines their values? It would be arrogant indeed to think that our generation of scientists has uncovered the "real" laws of nature, so the position is still unclear, but there are clues that an interrelatedness exists, waiting to be uncovered: these arbitrary numbers are either about one, or the huge number 10^{39}, surely this is not chance, there must be some relation between such huge numbers, but what is it?

THE PROBLEM

Our present description of the physical world is governed by empirically determined constants like the mass and charge of an electron, or the rate of expansion of the Universe, a partial list of these is

c	$= 3.0 \times 10^{10}\,\mathrm{cm\ sec^{-1}}$	:	velocity of light,
$\hbar$	$= 1.5 \times 10^{-27}\,\mathrm{erg\ sec}$	:	Dirac–Planck constant,
e	$= 4.8 \times 10^{-10}\,\mathrm{erg^{\frac{1}{2}}\ cm^{\frac{1}{2}}}$	:	charge on a proton,
m	$= 1.6 \times 10^{-24}\,\mathrm{g}$	:	mass of a proton,
m_e	$= 9.1 \times 10^{-28}\,\mathrm{g}$	:	mass of an electron,
G	$= 6.7 \times 10^{-8}\,\mathrm{erg\ cm\ g^{-2}}$	:	Newtonian constant of gravity,
H	$= 1.6 \times 10^{18}\,\mathrm{sec^{-1}}$	:	Hubble constant for the expansion of the Universe,
ρ	$= 4 \times 10^{-31}\,\mathrm{g\ cm^{-3}}$	:	mean density of the Universe.

These constants have dimensions, that is they are expressed in terms of some arbitrary chosen standard units, and so their numerical values have no significance, by changing the reference unit we change the numerical value of the constants of Nature. But from these constants we can form pure numbers, independent of any reference standard, these pure numbers contain the real empirical content of the laws of physics. From the constants listed above we can form the numbers

$$\frac{e^2}{\hbar c} = 7.3 \times 10^{-3} \qquad : \text{ fine structure constant,} \tag{1}$$

$$\frac{m_e}{m_p} = 0.54 \times 10^{-3} \qquad : \text{ mass ratio of electron and proton,} \tag{2}$$

$$\frac{e^2}{G m_p m_e} = 2.3 \times 10^{39} = C_1 \qquad : \begin{array}{l}\text{ratio of electrical to gravitational} \\ \text{force in a hydrogen atom,}\end{array} \tag{3}$$

$$\frac{m_e c^3}{e^2 H} \quad = \; 10.6 \times 10^{39} = C_2 \;: \quad \text{age of the Universe in atomic units,} \tag{4}$$

$$\frac{8\pi \rho c^3}{3 m_p H^3} \quad = \; 1.2 \times 10^{78} = C_3 \;: \quad \text{no. of particles in the observable} \atop \text{Universe.} \tag{5}$$

From the last three we can derive the result

$$\frac{8\pi}{3} \frac{G\rho}{H^2} = 0.08 = C_4. \tag{6}$$

The first three results are accurately determined, the others are less well known due to the uncertainty in the measurement of both the Hubble constant and the mean density of the Universe.

These numbers characterise the world we live in, if they had different values the Universe would be different, what are we to make of these results? The quantum physicist working in microphysics does not understand why the first two ratios have their measured values, but he believes that one day he will, these are not "God-given" arbitrary parameters but the only possible values they could have, it is just that as yet we do not know why. But what are we to make of the other constants, why are they so large, and why are (3) and (4) nearly equal to each other, and approximately the square root of (5)? Did God choose to set these constants equal to an arbitrary very large number like 10^{39}?

The Standard Explanation

The standard theory for "explaining" gravitation and cosmology is Einstein's General Theory of Relativity. In this theory the constant of gravity is God given, and as the Universe evolves so do the constants C_2 to C_4 change in time; in fact

$$C_1 = \text{constant}, \; C_2 \; \alpha \; f_1 \, (t), \; C_3 \; \alpha \; f_2 \, (t), \; C_4 \; \alpha \; f_3 \, (t) \tag{7}$$

where f_1, f_2 and f_3 are known functions of time that contain an arbitrary constant, God given. The observed coincidence

$$C_1 = C_2 = (C_3)^{\frac{1}{2}} \tag{8}$$

is then a transitory phenomena, we just happen to be living at the time when this is approximately true. The argument can be made more sophisticated, by arguing that in order for life to evolve and ask why these coincidences occur, the coincidences must occur! That is for $C_2 \ll C_1$ the Universe would not have evolved to the state where galaxies, stars, planets, and life would develop; for $C_2 \gg C_1$ the Universe would have evolved so far that there would no longer be stars providing the energy for life, only if $C_2 \simeq C_1$, would we expect life to exist, it is therefore not surprising that we observe $C_2 \simeq C_1$, if it were not so we would not be here![1]

This argument has several weaknesses: firstly, it requires the arbitrary constant that occurs in the functions $f_i(t)$ to be zero, or sufficiently small; secondly, it does not explain why C_1 has the particular value of 2.3×10^{39}. God just created the Universe with this value, it could have been different. But this is far from satisfactory, a complete theory of physics must explain why there exists the number 10^{39} in the description of the physical world, and why life evolved at all; the existence of life is a physical phenomenon that needs an explanation just like the existence of stars and galaxies. If conditions could have been different with a different choice of constants, why does the actual Universe have those values that give rise to life?

The alternative point of view is the same as that of the quantum physicist's view of the microscopic numbers (1) and (2). We do not yet know why there exists a number 10^{39} but it is knowable, it is not an arbitrary constant. This is the view advanced by Dirac[2] some years ago, and it leads naturally to the observation that as the age of the Universe in atomic units is 10^{39}, then the reason that $C_1 = C_2 = (C_3)^{\frac{1}{2}}$ now, is that they are interrelated; we may not as yet understand why, but as the only number of this magnitude is C_2 what else can be the explanation? But if this is correct now, it must also be correct in the future, and have been correct in the past, so that for all time

$$\frac{e^2}{Gm_p m_e} \simeq \frac{m_e c^3}{e^2 H} \simeq \left(\frac{8\pi\rho c^3}{3m_e H^3}\right)^{\frac{1}{2}}. \tag{9}$$

But this cannot be true with the standard theory since C_1 is constant whereas C_2 and C_3 vary in time, therefore the standard theory is wrong. But what is to replace it?

THE DIRAC COSMOLOGY

One way to proceed is to retain the standard representation of the expanding Universe according to which there is a standard Reimanian metric governing the behaviour of particles and photons

$$ds^2 = dt^2 - S^2(t)\left(\frac{dr^2}{(1-Ar^2)} + r^2 d\Omega^2\right). \tag{10}$$

This introduces another dimensionless number, the curvature of space at a given time, measured in atomic units

$$C_5 = \frac{Ae^4}{m^2 c^4} \tag{11}$$

which is necessarily constant in units where e, m and c are constant. We have no reliable estimate of C_5, it could be positive, negative or zero. For such a model we readily deduce that

$$H = \frac{\dot{S}}{S}, \qquad \rho = \rho_0 \frac{S_0^3}{\dot{S}} \tag{12}$$

and so

$$\frac{e^2}{Gm_e m_p} = \frac{mc^3 S}{e^2 \dot{S}} = \left(\frac{8\pi\rho_0 c^3}{3m_p}\right)^{\frac{1}{2}} \dot{S}^3. \tag{13}$$

and so

$$S(t) = S_0 t^{1/3}, \quad G \simeq \frac{e^4}{3m_p m_e^2 t}, \quad \rho = \frac{m_e^2 m_p c^3}{8\pi e^4 t}. \tag{14}$$

The constant of gravity G decreases with time as the Universe expands.

An Alternative Representation

General relativity is a theory of gravitation and inertia, the theory is complete without reference to the electrical properties of matter, this suggests that if we retain general relativity (or a modification of it) then it may be valid except with reference to electrical properties. Accordingly we solve the large number coincidences with G, m and c constant and find

$$G \approx \frac{1}{6\pi\rho t^2}, \quad S(t) = S_0 t^{2/3}, \quad e^2 \approx \frac{m_e c^3 t}{\left(\frac{4\pi}{3} n t^3 c^3\right)^{1/2}} \tag{15}$$

and since $\rho S^3(t)$ is a constant we have

$$\hbar c \, \alpha \, e^2 \, \alpha \, t^{1/2}. \tag{16}$$

As the Universe expands, the Dirac–Planck constant increases.

If these results can be successfully predicted by a theory then the number 10^{39} is nothing but the present age of the Universe, as the Universe evolves, all the large numbers vary in unison because they are just the same thing.

What about the constant C_5? It could be anything, it is another "God-given" constant. But if we accept that physics is interrelated and there are no arbitrary constants it must either be related to some existing number or be zero. It cannot be related to 10^{39} since C_5 is constant, the only possibility is that it is a small number of order unity. But this would give the radius of curvature of space of the order of 10^{13} cm, the size of an elementary particle. We therefore expect $C_5 = 0$, but our theory should force this conclusion.

MACH'S PRINCIPLE AND GRAVITATION

In the general theory of relativity the constant of gravity G is arbitrary — this is highly unsatisfactory, if we know the position and motion of all the particles in the Universe we can predict the future evolution without the introduction of an arbitrary constant. This concept (Mach's principle)[†] was the driving force behind Einstein's search for a relativistic theory of gravitation, but it failed, the theory does not make unique predictions. This is readily understood, Einstein's theory is a differential theory and differential equations have many

---

† This idea, though nowadays attributed to Ernst Mach, has its origins in the writings of Gottifried Leibniz and George Berkely criticising Newton's concept of absolute space.

solutions, if we wish to develop a unique theory we have to go to integral equations not differential ones. This has been recognised in recent years and in particular Hoyle and Narlikar[3] have developed theories along these lines. Can their theory be modified to provide the uniqueness we require? Technically the modified theory is given by

$$\delta \left(\int (\phi^2 R + 6g^{ij}\phi_{;i}\phi_{ij}) \sqrt{-g} \, d^4x + G \sum_i \int m_i \phi ds \right) = 0,\qquad(17)$$

$$\phi = \sum_i \int \widetilde{G}_- (\bar{x}, x) \, dx\qquad(18)$$

where G_- is the retarded Scalar greens function given by

$$\square_z \widetilde{G} (\bar{x}, x) + \frac{1}{6} R \, \widetilde{G}(z,x) = [-g(x)]^{-\frac{1}{2}}\delta^4 (z,x).\qquad(19)$$

This theory is formally equivalent to Einstein's theory in differential form, but as it contains an integral formulation of the mass field ϕ, not all the solutions of Einstein's equations are necessarily solutions of this theory, the integral acts as a filter. If it turns out that there is only one cosmological solution of the field equations — the Einstein de Sitter cosmology in which

$$6\pi \, G\rho t^2 = 1\qquad(20)$$

which for our purposes we interpret as

$$G = \frac{1}{6\pi\rho t^2}\qquad(21)$$

the constant gravity is determined by the cosmological distribution of matter. The other solutions of Einstein's equations, hyperbolic or elliptic universes, may not satisfy the integral formulation. C_5 is therefore zero, and the coincidences C_4 is satisfactorily explained.

MICROPHYSICS AND COSMOLOGY

It was Arthur Eddington[4] that drew attention to the uncertainty in physics caused by the Universe and its connection with microphysics. In making measurements in science we have to measure relative quantities, relative coordinates of the position of a particle and a reference frame, but what determines the reference frame? The best we can do is to take the centroid of all the particles in the Universe, or rather those within the horizon of an expanding Universe, this has the distinct advantage that the uncertainty in locating this centroid is known from the statistics of large numbers, without knowing the probability distribution of the individual particles. It is the Gaussian

$$f(x_0) = \frac{1}{\sqrt{2\pi\sigma^2}} e^{-x^2/2\sigma^2}.\qquad(22)$$

The uncertainty in position has an associated uncertainty in momentum, for a wave packet with a standard deviation σ, the distribution of momentum from standard wave mechanics is

$$\varpi(p) = \frac{1}{(2\pi\varpi^2)^{\frac{1}{2}}} e^{-p^2/2\varpi^2}, \varpi = \frac{\hbar}{2\sigma}. \tag{23}$$

What is σ? If we have N particles each of which can be anywhere in a sphere of radius R then

$$\sigma = \frac{R}{3\sqrt{N}}. \tag{24}$$

For a Universe full of photons moving with speed c, then at time t the distance to the horizon is $R \approx 2t$, and photons can have travelled a substantial fraction of R in the time t, so approximately

$$\sigma \approx \sqrt{\frac{ct}{N}}.$$

In the world there is an equivalence between mass and energy, can we understand mass as a concealed form of energy? If we took a model of the Universe with pure radiation and compared it with one of pure matter would we reveal this relationship between them? This was essentially Eddington's line of attack, we shall borrow his ideas but apply them to an expanding model universe (or Uranoid) comparing radiation- and matter-only models. This is quite straightforward using Einstein's equations in the mass field integral form we find in general

$$\text{matter: } 6\pi G \rho_m t_m^2 = 1,$$
$$\text{radiation: } \frac{32\pi G}{3} \rho_R t_R^2 = 1, \tag{25}$$

where ρ_m, t_m, and ρ_R, t_R are the density and age of the matter and radiation models. For the two models to have the same number of particles in the horizon of the same dimensions we have $t_R = 3t_m/2$, hence

$$\frac{\rho_m}{\rho_R} = 4. \tag{26}$$

Eddington in fact compared two static Einstein Universe and obtained 4/3 for this ratio.

What can we say about the mass density of radiation? The radiation-only uranoid is equivalent to the mass-only uranoid but with the particle velocities increased to c and the rest mass going to zero, i.e. an infinite temperature gas. But the momentum distribution and mean momentum of the particles is well behaved, since due to the fluctuations in the physical frame any momentum distribution relative to a geometrical frame is modified by the weight factor $\bar{\varpi}(p)$ which is the momentum distribution corresponding to the Gaussian distribution of position given in (22), thus the mean momentum of the photons is given by

$$\overline{p^2} = \frac{3\hbar^2}{4\sigma^2}. \tag{27}$$

But for a gas of photons the energy density is just the number of photons per unit volume n, multiplied by the mean energy per photon of $(p^z c^2)^{1/2}$, hence

$$\rho_{rad} = \sqrt{3} \, \frac{n\hbar}{c} \, \frac{1}{2\sigma}. \tag{28}$$

But for a radiation uranoid σ is $ct/\sqrt{N}$, hence

$$\rho_{rad} = \sqrt{3} \, \frac{n \, \hbar}{2 \, c^2 t_R} \sqrt{N} . \tag{29}$$

On the other hand, for the matter-only uranoid $\rho_m = nm_0$ where m_0 is the mass of the particle, where n is the same for the matter and radiation uranoids since N and R are the same, hence

$$\frac{\rho_m}{\rho_R} = 4 = \frac{2m_0 c^2 t_{rad}}{\sqrt{3N} \, \hbar} . \tag{30}$$

Now the number of particles in the horizon is $4\pi n c^3 t^3 /3$ and so using the constancy of the fine structure constant we find

$$e^2 = \frac{1}{137} \hbar c \simeq \left(\frac{m_0}{137 m_e} \right) \frac{m_e c^3 t}{[4\pi n t^3 c^3 /3]^{1/2}} \tag{31}$$

which is just the relation (15) provided m_0 is of the order of the pion mass.

We cannot hope to do any better than this at this stage, we would hope like Eddington that m_e and m_p are determined in terms of this mass constant m_0, but how is still a mystery.

CONCLUSIONS

The preliminary results obtained by this simple analysis are encouraging, they give the correct dependence of G and e on cosmic time t, they give the correct order of magnitude for the cosmic numbers. 10^{39} is the current age of the Universe and the other numbers of this magnitude are causally related to the age. Of course we have not solved all problems, we need to show why in this theory the Universe has the properties it has, galaxies, stars, planets and people at the present time of 10^{39}. This is the same problem as in the standard theory, and a similar reasoning, with similar uncertainty would lead to the conclusion that life emerges when $C_2 \approx 10^{39}$.

But there are also empirical checks on the theory, if as measured in atomic units e, m, c, the constant of gravity changes $G \sim 1/t$, then the properties of stars and the orbit of the Moon round the Earth or the Earth round the Sun should show a secular change. This change would be of the order of 1 part in 10^{10} years and some indication of this has been suggested by the work of van Flandern.[5] It can also be measured in the laboratory and an experiment is currently underway at the University of Virginia to look for this. If it is found it will strongly support the contention that the Universe is unique, we may not have a full theory for the interrelatedness of nature, the connection between cosmology and microphysics, but perhaps we are on the right lines.

REFERENCES

1. R. Dicke, *Nature*, **192**, 440 (1961); B. Carter, *Proc. I.A.U. Symposium* no. 63, p. 291 (1974).
2. P. A. M. Dirac, *Nature*, **165**, 199 (1937).
3. F. Hoyle and J. Narlikar, *Proc. Roy. Soc.* A, **282**, 191 (1964).
4. A. S. Eddington, *Fundamental Theory*, C.U.P. (1966).
5. T. C. van Flandern, *Science*, p. 44 (1976).

William H. McCrea, F.R.S.

Educated at Trinity College, Cambridge, and Göttingen University;
former Bye-Fellow of Gonville and Caius College, Cambridge; Professor
of Mathematics in Queen's University, Belfast, 1936-44 and in London
University (Royal Holloway College) 1944-66; Professor of Astronomy,
Sussex University 1966-72, Emeritus since 1972. Past President of the
Mathematical Association and the Royal Astronomical Society.
Awarded the Royal Astronomical Society Gold Medal 1976. Research
in mathematics, mathematical physics, astrophysics and cosmology
including problems of the origin and evolution of the solar system and
particularly the significance for such problems of the large-scale
structure of the Galaxy.

ORIGIN OF EARTH, MOON
AND PLANETS

The problem of the origin of the solar system is perhaps the most notable of all unsolved problems in astronomy. It has unique relevance to ourselves and our place in the cosmos; it ought to be solvable in the present state of general astronomical knowledge; the correct solution, if found, should be recognizable as correct; numerous attempts to solve it have, in fact, been made; no problem of the formation of *any* astronomical body has ever yet received an accepted solution. All other astronomical problems lack some of these features. There are, for example, problems of origins of much greater magnitude but they are of less immediate concern to ourselves; also the greater the scope the more difficult it is to formulate a problem, and the less certain that a solution may be achievable or even recognizable. Elsewhere (McCrea, 1977) I have tried briefly to describe some wider aspects of the astronomical setting of the problem. In the present article I seek to discuss the problem on its merits.

SOLAR SYSTEM

The composition of the solar system is summarized in Table 1. There is the Sun, a rather

TABLE 1. Solar System

	Mass or mass-range Earth-masses	Mean density or density-range, g/cm^3
Sun	3.3×10^5	1.4
Terrestrial planets: Mercury Venus, Earth, Mars, Pluto	0.06 to 1.0	4 to 5.5
Major planets: Jupiter, Saturn	95 to 320	0.7 to 1.3
Outer planets: Uranus, Neptune	15 to 17	1.6 to 2.3
Asteroids (minor planets)	5×10^{-4} total mass	
Interplanetary gas, dust, meteoroids, comets		

$$\frac{\text{Mass of planetary system}}{\text{Mass of Sun}} \sim \frac{1}{700} \qquad \frac{\text{Angular momentum of planetary system}}{\text{Angular momentum of Sun}} \sim 200.$$

Planet	Known natural satellites		Miscellaneous satellites	Angular momentum of main satellites
	Regular satellites			
	Main	Minor		Angular momentum of planet
Mercury	—	—	—	—
Venus	—	—	—	—
Earth	Moon	—	—	5
Mars	—	—	2	—
Jupiter	Io, Europa, Ganymede, Callisto	1	8	1/100
Saturn	Titan	6 + rings	3	1/80
Uranus	—	5 + rings	—	—
Neptune	Triton	—	1	~1/20
Pluto	—	—	—	—

Total mass of main satellites ~ 0.16 Earth mass.

average normal star, which is 1/3 million times as massive as the Earth; there are nine planets, the heaviest, Jupiter, being nearly 6000 times as massive as the lightest, Mercury, and the densest, the Earth, has nearly 8 times the mean density of the least dense, Saturn. There are thirty-three known natural satellites shared between six of the planets. There are much smaller bodies in considerable variety and numbers. Proceeding in this way, the range of properties is bewildering. It becomes necessary therefore to ask what, if any, basic regularities may be discerned. Fortunately, it is possible to recognize quite a number of these, as follows:

Mass and composition. The six principal planets Venus, Earth, Jupiter, Saturn, Uranus, Neptune *all have roughly the same heavy-element content*, which is therefore a standard feature. One way of looking at this is that Venus and Earth are very similar in mass and mean density; bodies similar to the other four of these planets could be got simply by taking Earth or Venus and surrounding it by suitable quantities of the lightest elements hydrogen and helium. A probably more significant description is this: the composition by mass of the Sun is about 70 per cent hydrogen, 28 per cent helium, 2 per cent heavier elements; of all the planets Jupiter has most nearly the same composition; were perhaps 50 per cent of the light elements hydrogen and helium removed from Jupiter, the result would be quite like Saturn; were some 90 per cent removed the result would be like Uranus and Neptune; were 99 per cent removed, the result would be like Earth and Venus; in fact, the principal planets are all derivable from six proto-planets all having the same chemical composition and not far from the same mass. The "lesser planets" Mercury, Mars, Pluto are bodies of chemical composition somewhat like Earth and Venus, but of the order of only one-tenth the mass.

Rotation. The spin-axes of the planets are inclined to the orbital axes at mostly quite considerable angles. In two or three cases the present rotation-rate is obviously subject to the tidal influence of other bodies in the system. However, in six cases the rotation-period is between about 10 and 25 hours; that is to say, *we may recognize a roughly standard natural rotation-period* with a median of, say, 15 hours.

Revolution. All the planets revolve about the Sun in the same sense in orbital planes that are within a few degrees of each other, with fairly small orbital eccentricities. Roughly speaking, each orbit has getting on for double the size of the next smaller orbit; were a theory to insist upon a totally different arrangement of sizes it would be unacceptable; it cannot be claimed, however, that there is a regularity of sizes ("Titius-Bode formula") to which a cosmogonic theory as such must unquestionably conform. Incidentally, the Earth's orbital plane is not the equatorial plane of the Sun, but is inclined to it at about 7 degrees.

Satellites. The six main satellites of Earth, Jupiter and Saturn, even as they exist, *are the most standard bodies in the solar system* — the satellite Triton of Neptune is probably in this category, but values of its parameters are still somewhat uncertain and it is best to leave it out of the discussion; three of them have about the same density as the Moon, which is believed to be composed mainly of silicates; it is inferred that the other three have mantles (or admixtures) of ices, and that the silicate portions of all six are even more standard bodies. These main satellites move in orbits about the planets that have small, but not all negligible, eccentricities, and whose axes in two cases have large inclinations to the rotation axes of the planets.

The twelve very much smaller satellites classed as "minor" are all within 10^5 and 6×10^5 km from the three relevant planets, round which they revolve in orbits of very small eccentricity and very small inclination. In the three cases in which mean densities have been well estimated the values lie in the range 1.1 to 1.3 g/cm^3; the material is probably the same as the mantle-material (when present) of the main satellites. Saturn's rings appear to be closely related to its inner (minor) satellites. *The evidence is that the minor satellites are products of a standard process.*

The fourteen miscellaneous known satellites are all very small, and they are clearly distinct from regular satellites — although maybe some were formed along with the minor satellites and suffered subsequent disturbance. The fact that *three planets have no known satellites* must be significant.

Finally, in the solar system as a whole *nearly all the angular momentum is associated with a very small part of the mass*; in the satellite systems the reverse is generally the case.

The main regularities are those in the italicized statements, they strongly indicate that the solar system was formed as the one system and that we should look for a theory of its origin as such. The principal requirement of a theoretical model is to reproduce these features. At the same time, it has got to reproduce them as they actually occur as *approximate* regularities — a theory that required, say, all the planets to have orbits precisely in one plane with precisely zero eccentricity might be more objectionable than one making no particular prediction about these characteristics.

We know only the one planetary system, and this is obviously a chief source of difficulty in discovering its origin. Within that system, however, we know six satellite systems — or more significantly nine, of which three are almost certainly void. Therefore one has a better chance of discovering the origin of a satellite system than that of a planetary system. But the formation of satellite systems is obviously closely bound up with that of the associated planets, so anything learned about satellite formation should throw light upon planet formation. This is why it is so important to take account of satellites, even though they be lightweight members of the system as a whole. At the same time various properties, particularly in regard to angular momentum, show that a satellite system is *not* like a planetary system in miniature. Indeed, if a model that has been proposed for planet-formation should turn out to be good for satellite-formation, then that probably shows it to be bad for planet-formation!

As regards the possibility of observing another planetary system, planets like ours associated with any other star like our Sun would be utterly undetectable by any available means. Van de Kamp (1975) has collected a number of cases where a less massive star shows variability of motion possibly attributable to the presence of a specially massive planet, but nothing is yet known that helps the present study.

RAW MATERIAL

Current models in general postulate raw material having about "solar" composition. At the relevant time it has to be cold ($\lesssim 100°$ K); at such a temperature and at the relevant density, much of the heavier-element content would have condensed to form minute grains of silicates, metals, graphite, ices, to the extent of about 1 per cent of the material by mass. The remainder would still be in the gaseous state and would consist mostly of hydrogen and helium, with about another 1 per cent of heavier elements not incorporated in the grains.

Were all the material of even the entire solar system (say 2×10^{33} g) spread evenly through the volume of a sphere with radius about the distance of Neptune (say 5×10^{14} cm) the density would be only about 4×10^{-12} g/cm^3. The processes to be considered must take place through a region of some such size, and they cannot involve much more than this amount of matter; therefore on any model the initial mean density of the raw material is extremely low — less than a good terrestrial vacuum.

The most significant way of classifying models is into those that treat the Sun and the requisite raw material in isolation, and those that consider them necessarily as part of some larger assembly.

Solar nebula. A model in the first of these classes always reaches a stage when the Sun, either while in course of formation or sometime later, is surrounded by a well-formed *solar nebula*. This is a body of the raw material having mass-distribution and momentum-distribution such that when some subsequently adopted process is inferred to form the planets from it they will turn out to have the correct masses and motions.

One proposed mechanism for producing the nebular material is that the gravitational attraction of the Sun would entrap it after drawing it out by tidal action from a passing diffuse star. In recent years, Woolfson (1969) and his co-workers have shown by computation that this is dynamically feasible; the basic suggestion is much older.

Another mechanism is for the material to be a residue from the formation of the Sun from an original a solar nebula — as long ago proposed by Laplace — after there has been outward transfer of spin-momentum, say by magnetic coupling, or after considerable loss of mass without too great loss of spin, as proposed by A. G. W. Cameron (1976) and his group.

Finally, as the Sun travels through interstellar clouds it may capture cloud-material in its gravitational field and such material may settle down into orderly rotation around the Sun. In this context, the process was first considered many years ago by Schmidt (1959).

Star formation and protoplanets. An important part of the Milky Way Galaxy consists of stars and clouds of gas and dust that form a disc, and that a remote observer would see as a spiral galaxy, like the Andromeda Galaxy. The stars and clouds are all in orbit round the galactic centre, producing the phenomenon of "galactic rotation". The motion as a whole is such that there exist regions where there is a traffic-jam; these regions *are* the spiral arms. Clouds running into a jam sometimes suffer a shock which causes them to be considerably compressed, as readily seen in the features known as "dust-lanes" in photographs of other galaxies (Lynds, 1974). Clouds normally recover as they emerge from the jam; occasionally, however, a cloud gets so compressed that gravitational effects within it overcome its tendency to re-expand, and instead it condenses to form a cluster of new stars. This also is seen in our own and other galaxies in that what are recognized as young clusters or young stars are in fact observed in the appropriate locations. All this is quite direct inference from much observation and it involves no speculation.

Astronomers are convinced that they are witnessing the formation of new stars in certain parts of the sky; the phenomena are compatible with the foregoing general picture. Unfortunately there is still no agreed model for the way in which a cloud of interstellar material does become resolved into stars. It seems simple to say that a cloud breaks into fragments and each fragment contracts under its self-gravitation; the trouble is that by the time it had contracted to the density of a star it would be spinning with equatorial speed exceeding the speed of light! The escape from this difficulty must have to do with the fact that a lot of stars are formed all at once. It may be that something very drastic happens to the cloud as a whole before the stars are formed from it. Or it may be that at a critical stage the cloud as a whole "clots" or "coagulates" into stars; the picture then would be that blobs of cloud-material coalesce here and there into what become centres of gravitational condensation leading primarily to stars; after these main condensations have been formed, there would be some residue of blobs that have too much angular momentum about every one of the main condensations for them to fall into main condensations — some will disperse and some may just manage to hold together and survive as what we shall call *protoplanets*. These will be shared amongst the gravitational fields of the newly formed stars, or at any rate such of those stars as possess no stellar companion. In the process of assembling any protoplanetary system, it will tend to flatten out, in the fashion that, for instance, the particles of Saturn's rings have settled into a plane, but on a coarser scale in the case of protoplanets (McCrea, 1960). Here the point is that when we recognize that stars are formed in clusters, it is more natural to infer that some of them acquire a family of protoplanets, rather than a solar nebula.

PLANET FORMATION

Once the raw material has been deposited around the Sun in the shape of either a solar nebula or protoplanets there are two possibilities for the formation of normal planets from it:

Accumulation. Most models that invoke a solar nebula envisage planets are being produced from it by a process of accretion or *accumulation*. Some such models give most attention to mechanical, thermodynamical and thermochemical processes taking place before the appearance of the planets, with the object of specifying a nebula that is self-consistent in regard to its density distribution and gas dynamics, its heating and cooling at any point, the condensation of grains, the evaporation of light gases across its boundary, and so forth. If it be then accepted as self-evident that planets will indeed form from belts of this nebula, there seems to be no doubt that the dependence of chemical composition of actual planets upon distance from the Sun can be reproduced in a convincing manner. But if the formation of planets be thus taken for granted, then the model is not asked to predict the number, sizes, spins, satellite-features, etc., of these planets.

Of course nobody really takes it to be self-evident that planets must be formed at all under the inferred conditions, and so some models give major attention to this aspect. For example, Dole (1970) has carried out extensive computer simulations of the formation of planets, according to a stated prescription, in a solar nebula specified by several adjustable parameters. With particular values of these parameters, he succeeded in reproducing the actual planetary system with astonishing fidelity. Rather reluctantly, one has to state three main criticisms:

(i) The solar nebula described by Dole's parameters is not shown to be self-consistent in the above-mentioned sense — so far as one knows, nobody has married a fully worked out process of forming planets to a fully worked out structure of the solar nebula.

(ii) Dole's computations expressly give no information about the duration of the processes; it is quite unknown whether the postulated nebula would endure long enough for the processes to attain consummation.

(iii) Dole's processes would not get off the ground — to venture an incongruous use of the current idiom — without his postulated nuclei for planet-accumulation which are an arbitrary element in his prescription.

In order for the accumulation ever to get started, the view developed in recent years

(Goldreich and Ward 1973) is that after the solar nebula has reached the state already
mentioned, yet another process supervenes — under the gravitational field of the nebular
material itself, the grains fall through the gas and settle into a layer in its equatorial plane.
Accidental irregularities in this layer, if sufficiently extensive, will contract gravitationally upon
themselves; condensations initiated in this way are then considered to set in motion a train of
processes such as that treated by Dole. Those who follow a theory like this usually claim that
satellites arise by a further application of the theory.

Protoplanet evolution. As we have seen, protoplanets seem to arise naturally in the context
of the formation of a stellar cluster, although they *might* also arise by fragmentation of a solar
nebula. Discussion shows that, by whatever means it does arise, a protoplanet would be initially
a rather standard body of the raw material, of about the mass of Jupiter, in a cold diffuse state
but sufficiently dense to hold itself together by its self-gravitation; it is in orbit round the Sun,
but it will evolve largely as an independent body. Like any naturally occurring body it has
non-zero spin-momentum about its mass-centre, and this is conserved according to ordinary
mechanical principles.

The diffuse protoplanet will proceed to contract under its own gravitation, the gravitational
energy released being radiated away so long as the material is sufficiently transparent. But as
the size diminishes the body inevitably spins faster; ignoring for the moment rotational
distortion, at some radius R it will start spinning off material round its equator. The body will
go on contracting and leaving material behind in the equatorial plane until its remaining
material becomes relatively incompressible. For a mass like Jupiter's this must occur at mean
density about that of Jupiter, i.e. about 1 g/cm^3. It is still rotationally unstable, but for
material of small compressibility this will be relieved once and for all by fissional break-up in
which the main body casts off a portion about one-tenth its mass. Thus there would be, as
calculation shows, about half of the original material near the equatorial plane out to distance
R and most of the rest of the mass forming a stable planet in the middle. *This is the proposed
origin of Jupiter and Saturn.*

These ideas go back more than 50 years to the work of Sir James Jeans which has in part
been redeveloped more recently by Lyttleton (1960). More modern work has been done on the
whole subject but most of it is applicable rather to the rotational flattening and possible
break-up of a star than to a planet or protoplanet. One can either accept that the treatment
outlined above is valid for the present application or one can regard it as a rough approximation
to what would be expected from a theory along modern lines adapted to this application.

Part of the story is that *the material left near the equatorial plane is that from which
(regular) minor satellites are formed* in this model. The process would be the accumulation
process envisaged in the preceding section as a process for forming planets from a solar nebula;
here we appeal to it on a much smaller scale and we require only very low efficiency; it would
explain why these satellites have (closely) circular orbits in the equatorial plane of the planet
concerned. Also on this interpretation R is about the outer radius of the set of these satellites in
each case and this serves to set a value on the original spin, and this in turn determines how
much of the original mass is left to form the main resulting planet. On the model the final spin
is significantly less than the critical spin that caused rotational break up, but still of that order,
which implies a rotation period of order 10 hours as actually found.

There is, however, another process that to a greater or lesser extent must operate from the
outset: the grain material must tend to sediment towards the centre of the protoplanet. In the
solar nebula model the same process causes this material to settle towards the central plane, but

it would be more effectual in a protoplanet. The evolution of a protoplanet as already described applies if the sedimentation rate is slow compared with the contraction rate. However, we may consider the other extreme case in which the light gases continue to occupy about the original volume of the protoplanet while only the heavier material contracts towards the centre. Much the same treatment as before applies now to *this* material. But there are two adjustments: as the heavier material contracts through the excess of lighter material it leaves behind most of its spin-momentum; and the density at which the contracting material becomes effectively incompressible, and break-up occurs, is greater than before — presumably about the mean density of the Earth which is composed of just such material. *On the model the surviving main body is indeed the Earth or Venus.*

We have now rather literally to tidy up this model. In the case just considered the diffuse gas and some discarded grain-material are left behind. If the protoplanet is near enough to the Sun such material would be removed by tidal action and by the solar wind. On the model there would therefore be no possibility of the occurrence of (regular) minor satellites of a terrestrial planet. On the other hand, if the protoplanet is far enough from the Sun some diffuse gas and any other material left beind in the sedimentation process may proceed to contract as in the process first considered. *This would be the present explanation of Uranus and Neptune* and it would allow them to have minor satellites — Uranus actually has them, Neptune has no regular minor satellite.

If a body in orbit about the Sun undergoes rotational fission the main portion remains in a not much different orbit; Lyttleton has shown that the smaller portion goes into a separate orbit if the original orbit is at about the Earth's distance from the Sun or nearer, while it would escape altogether if the original orbit is at about Jupiter's distance or further. The implication is that *Mars is the smaller portion from the fission of proto-Earth, and Mercury that from proto-Venus.* The inferred fission of proto-Jupiter and proto-Saturn would have produced no surviving smaller planets. Lyttleton's conclusions apply, however, only if no other matter intervenes; if the lighter material of proto-Neptune is retained as suggested, this will retard the escape of the smaller fission portion which then need not be lost to the solar system. *The implication is that such was the origin of Pluto.*

When a fluid body is on the point of fissional break-up into two portions, these are connected by a "neck" of this fluid. Fission is the demolition of this connection; directly it happens some of the connecting material may fall back into the two main bodies, but some will form "droplets" strung out between them. The description is essentially Lyttleton's; he has shown how some droplets would be expected to go into orbit about the larger main portion, some would escape, none could go into orbit about the smaller main portion because any setting out to do so would collide with that body (collision fragments might survive as tiny satellites). A suggestion apparently due originally to Lyttleton is that *the Moon may be such a droplet.* If it is, then *all seven main satellites would have the same sort of origin* and it is satisfactory that the model requires fission in all relevant cases. Venus might have had a moon in this way, but owing to solar tidal friction its day has become longer than its month could be and so "lunar" tidal friction would cause a moon to fall into the planet. Finally, various escaping bodies have been mentioned; had any of these suffered collisions amongst themselves then there would have been plenty of fragments to account for asteroids and meteorites, and may be also miscellaneous satellites.

The protoplanet model appears therefore to account for the main regularities in the solar system and to provide for the variety of its properties both positive and negative — it explains, for example, not only why certain planets have satellites of certain sorts but also why others do not have them.

The model qualifies nevertheless, along with the rest, for inclusion in an encyclopaedia of ignorance. One cannot claim that things had to be like this.

DISCUSSION

There are compensations in writing about ignorance: there can be no completeness, and no author is likely to complain if one fails to quote his particular contribution to the sum total of our ignorance. However, there are embarrassments: if ignorance is complete, there is nothing to be said about it anyhow, and if knowledge is fairly complete it does not qualify for this book. So any discussion that does qualify necessarily concerns the in-between region — that which appears to be just ahead of the edge of knowledge. And here it is that one's personal preferences and expectations have most influence. So one can only try to make clear why one has adopted a particular presentation, while freely allowing it to be biased.

My first concern is to show why there is such profusion of ignorance about the origin of the solar system. It is primarily because the subject is so complex, and I thought I could demonstrate this only by trying to show the reader how people are actually trying to tackle it — if possible to give a *feeling* for the subject. I hope this brings out the difference from ignorance in other fields; in pure mathematics, for instance, much ignorance may depend upon failure to prove or disprove one single theorem; in biology, progress in some branch may be held up because some quite specific question in chemistry is still unanswered; there are problems in the foundations of physics where no one knows even how to begin, and so forth. Nevertheless it might be more true to say, not that ignorance in cosmogony is different, but that it includes all possible kinds of ignorance encountered elsewhere. And for that very reason, as should be apparent from the account given in this paper, the problem of the origin of the solar system gives rise to a series of problems each of which is of a more usual scientific sort. Happily considerable progress is being made in these individual problems. There is even considerable agreement that the results are indeed those needed for solving the overall cosmogonic problem. Where there is most uncertainty and disagreement is in the way in which those results can be combined into a coherent whole.

A minor but still significant question is the role of tradition; this happens to be well illustrated here. Interest in the origin of the solar system goes back to long before modern ideas on the structure of the Galaxy, the behaviour of interstellar clouds and the formation of stellar clusters. If raw material from interstellar space was out of the reckoning, then it could be got only out of a star or as a residue in the making of the Sun; this led to the two traditional approaches. As indicated in this paper, they still provide the starting places of much current work. It is pardonable to ask to what extent this is simply because even in science, tradition dies hard.

REFERENCES

Cameron, A. G. W. and Pollack, J. B. (1976) in *Jupiter*, ed. T. Gehrels (University of Arizona Press).
Dole, S. H. (1970) *Icarus* 13, 494-508.
Goldreich, P. and Ward, W. R. (1973) *Astrophys. J.* 183, 1051-61.

Lynds, B. T. (1974) An atlas of dust and H II regions in galaxies. *Astrophys. J.* Supplement number 267.
Lyttleton, R. A. (1960) *Monthly Not. Roy. Astron. Soc.* **121**, 551-69.
McCrea, W. H. (1960) *Proc. Royal Soc.* **A256**, 245-66.
McCrea, W. H. (1977) in *Origin of the Solar System,* ed. S. F. Dermott (John Wiley).
Schmidt, O. (1959) *A Theory of the Origin of the Earth; Four Lectures*, London, Lawrence and Wishart.
van de Kamp, P. (1975) *Ann. Rev. Astron. and Astrophys.* **13**, 295-333.
Woolfson, M. M. (1969) *Progress Phys.* **32** 135-185.

General Reference

Dermott, S. F. (ed.) (1977) *Origin of Solar System*, John Wiley.
Reeves, H. (ed.) (1972) *Symposium on the Origin of the Solar System*, Paris, CNRS.

Michael Rowan-Robinson

Lecturer in Mathematics at Queen Mary College, University of London.

Main interest has been the application of the "new" astronomies, particularly radio and X-ray astronomy, to cosmology. A special interest has been quasars, their distances and evolutionary properties. Increasingly involved since 1974 with far infra-red astronomy, working with the Queen Mary College far infra-red group at Kitt Peak, Arizona.

GALAXIES, QUASARS AND THE UNIVERSE

In extragalactic astrophysics we suffer from three types of ignorance: ignorance in principle, ignorance due to observational limitations, and ignorance due to inadequacy of theory and observation.

An example of ignorance in principle is associated with the finite time that light takes to reach us from distant galaxies. While in our locality we have evidence about our remote past, for distant galaxies from which the light set off long ago we have no evidence about how they look *now*. We can thus never test one of the most basic philosophical ideas underlying modern cosmological theory, that the universe is *homogeneous*, each observer seeing the same picture, on average, as we do. There is a considerable logical gulf between the earth not being the centre of the universe and perfect homogeneity. Virtually no theories populate this gulf, although the universe itself clearly does. The sceptic (a rare bird among cosmologists) would like to see a model of the universe less dependent on metaphysical abstractions like the principle of homogeneity. It is true that perfect *isotropy*, together with the assumption that we are not in a special place, implies homogeneity. Unfortunately, the distribution of matter in the universe is certainly not isotropic. Just look up at the night sky! Can the lumpiness of the matter distribution really be ignored?

We cannot count observational limitations due to the earth's environment as sources of ignorance in the long term, for surely most of these can be overcome, if we so desire. The earth's ionosphere reflects all radio waves of frequency $\lesssim 30$ MHz, but lower frequency radiation can be detected from satellites. Virtually no radiation with wavelengths between 20 μm and 1 mm reaches sea-level, but dry high-altitude sites, aircraft, balloon (and soon satellite) observations are opening up the far infra-red and sub-millimetre bands for us. No cosmic X-rays reach the surface of the earth, but rocket and satellite experiments have already mapped the X-ray sky. The angular resolution of ground-based optical telescopes is limited by atmospheric fluctuations to $1''$ arc, but this can be overcome by balloon or satellite observations. Indeed at any time there will be a finite achievable sensitivity, angular resolution, spectral and time resolution at each wavelength, the transcending of which will lead to new detail in our knowledge and occasionally to entirely new types of phenomena.

However, there are some genuine observational limitations which define areas of ignorance for the forseeable future. The very lowest energy radio waves ($\lesssim 10^4$ Hz) and the highest energy γ-rays ($\gtrsim 10^{24}$ Hz) cannot reach us from outside our own Galaxy. The interstellar medium restricts our horizon in far ultraviolet light to only about 100 light years, compared with the 10^{10} lt.yr or more accessible in the visible. The absorbing and scattering power of minute grains of dust in the interstellar medium hides from our view most of the plane of the Milky Way, and the cores of dense clouds where stars are forming, except to radio and infra-red waves. But when we consider how the electromagnetic spectrum has been opened up for astronomy in the past 30 years, these areas of ignorance seem to be continually dwindling.

It is when we turn to the third type of ignorance, that due to the inadequacy of theory and observation, that we have greatest cause for modesty in our claims of achievement in astronomy. True we have some sort of overall scenario for the evolution of matter in the universe, and this represents a great advance on the situation as recently as a decade ago. It is, however, a scenario riddled with untested assumptions.

Firstly we assume, with no evidence, that Einstein's General Theory of Relativity applies to matter on the large scale. An associated assumption is that the gravitational constant and other physical constants do not change with time. Then we assume that matter is distributed homogeneously and isotropically through the universe. We assume, again without evidence, that the redshifting of the spectral lines of galaxies, which is found to be proportional to distance (the Hubble law), is due to the expansion of the universe.

The picture is that about 2×10^{10} years ago the universe exploded out from a state of infinite density, temperature and pressure (the Big Bang). For the first 10^6 years the dominant form of energy was radiation, with the less significant matter in thermal equilibrium with it. The temperature fell as the ratio by which the universe expanded and at around 10^9 K nuclear reactions ceased, leaving protons, electrons, neutrinos, about 27 per cent helium by mass, about 10^{-5} deuterium (^{2}H), and traces of other elements.

When the temperature dropped to about 3000 K, after 10^6 years, protons and electrons recombined to form neutral hydrogen and the universe became suddenly transparent, allowing matter and radiation to decouple. The radiation travelled freely through the universe until, redshifted by a further factor of 1000, it reaches the observer today as a microwave background radiation with the spectrum of a 2.7 K blackbody. The matter, on the other hand, was now free to condense into galaxies, presuming the basically uniform gas to be dimpled with protogalactic irregularities in density. After about 10^9 years the collapse of the fragments was complete. Some formed almost entirely into stars as they collapsed, yielding elliptical galaxies. Others formed only partly into stars, the residual gas forming a rotating disc: spiral density waves rotate through the disc triggering star formation in their wake to yield spiral galaxies like our own.

Both types of galaxy form a dense core, or nucleus, which may be continually fuelled by gas ejected from evolving stars or supernovae (the final explosive stage of massive stars). In this nucleus a dense star cluster, supermassive star or giant black hole, forms from time to time, acting as a powerful accelerator of relativistic particles. These give rise to a variety of types of "active" galaxy: radio-galaxies with strong sources of non-thermal radiation centred on the galaxy or located in double sources far from the galaxy, galaxies with non-thermal optical cores (Seyferts, N-galaxies or, when the core outshines the whole optical galaxy, quasars).

This then is the framework within which we can understand most of what we know about galaxies, quasars and the universe. Yet there are so many gaps in the picture, so many areas of doubt or controversy, that the sceptic asks if it is really a picture with any predictive power at all.

The first question, and in many ways the most crucial for the whole Big Bang picture, is whether the 2.7 K blackbody background radiation is indeed the relic of an opaque, radiation-dominated phase of the universe. Recent balloon-borne experiments have shown that the spectrum does indeed start to turn down at wavelengths short of 1 mm, as expected for a blackbody spectrum. The spectrum and the great smoothness of the radiation (isotropy on the small scale) make it extremely difficult to suppose that it is due to a superposition of faint discrete sources. The high degree of isotropy on the large scale (± 0.1 per cent) is the only solid evidence for isotropic (and homogeneous) models of the universe and soon the local motion of the solar system should be clearly measurable, an experiment as fundamental as that of Michelson and Morley. One disquieting fact about this background is that its energy-density is rather similar to that expected from the integrated starlight from galaxies, but no satisfactory method of thermalizing this radiation to give a 2.7 K blackbody has been devised. If nature has not found a way, then we simply happen to be alive at the epoch of this coincidence. The neo-Aristotelians might counter with the "anthropic principle": the universe is as it is because if it were not, we would not be here to observe it. Still more inexplicable is the coincidence between the energy-density of the microwave background and those due to cosmic rays, and also to magnetic field, in our Galaxy.

Another important plank in the Big Bang platform is the "primordial" abundance of helium, found in the oldest stars in our Galaxy. While this could have been formed in an early, long-vanished generation of stars (and such an early burst of star-formation can also be argued for from the absence of stars with very low abundances of heavier elements which *have* to be formed in stars), it is generally assumed to be another relic of the "fireball" phase. *Ad hoc* explanations are needed for one or two stars in our Galaxy in which *no* helium is seen in their surface layers. Another accepted candidate for cosmological honours is deuterium since though it is easily destroyed in stellar interiors, it is hard to make. The observed abundance requires the average density of the universe now to be about 10^{-31} g cm^{-3}, implying that the retarding effect of gravity on the expansion of the universe is almost negligible. This is also about the observed average density of matter in galaxies.

Before we leave those questions directly connected with the fireball phase of the Big Bang, we have to ask why we live in a universe in which matter dominates over antimatter. For in the conventional picture, matter and antimatter must have coexisted in thermal equilibrium in almost identical amounts when $T > 10^{12}$ K. Why was there that slight excess of protons and electrons over antiprotons and positrons to survive annihilation? Do I hear the anthropic principle being wheeled out again?

The formation of galaxies provides one of the thorniest problems in cosmology today. Despite intensive work, no solution has been produced which does not amount to saying: a galaxy forms because the initial conditions of the universe preordained that it would. Significant density fluctuations are an essential feature of the creation. The same applies to the formation of clusters of galaxies. This seems rather a devastating blow to the philosophy that the universe started off simple and structureless, and then subsequently developed the structure that we see. An interesting development of recent years has been the idea that the universe started off extremely *in*homogeneous and *an*isotropic, and then evolved towards isotropy and homogeneity, retaining only those irregularities needed to form galaxies and clusters. A totally different approach is to suggest that the nuclei of galaxies and quasars are sources of continuous, spontaneous creation of matter, and that galaxies grow rather than condense. Such ideas were more popular when the steady-state cosmology, which brought the unmentionable act of creation into the astrophysical arena, was in vogue.

There is a real paradox associated with rich clusters of galaxies, that they do not seem to be gravitationally bound by the matter we see in them. The velocities of the galaxies ought to be fast enough for them to escape from the cluster, causing its dissolution. Could there be enough invisible or undetected matter, e.g. planets, isolated stars or star-clusters, dwarf galaxies, intergalactic gas, or for that matter rocks and burnt-out space-ships, to bind the cluster? The most popular candidate (on the basis that galaxy formation cannot be 100 per cent efficient) is residual intergalactic gas. There is evidence for such gas, both from its X-ray emission and from its effect on radio-sources, but not in the required quantities. This leads naturally to the question of a general intercluster gas. Such gas would have to be ionized and at a temperature of about 10^6 K to avoid detection to date. If deuterium is made in the fireball, the density of intercluster gas cannot be much above 10^{31} g cm^{-3}.

Stars – clusters of stars – galaxies – clusters of galaxies: does the hierarchy end there? While complexes consisting of several clusters of galaxies are known, there is no evidence that this is a general feature of the cluster distribution. Yet although we can see galaxies at distances of up to 10^{10} lt.yr, so that we are looking backwards in time at least half-way to the Big Bang, the *spatial* fraction of the universe explored so far by observation is almost certainly negligible. If what we have seen so far is only a chunk of some Metagalaxy, then all we can say is that other metagalaxies must be very distant, to avoid too great an anisotropy in the radio source distribution.

Surely when we turn from the large scale to our own Galaxy, we must be on more solid ground. But even here we are faced with at least one immense lacuna in our knowledge, the problem of star formation. By mapping out the location of the youngest stars we see that they are found preferentially in the spiral arms. These arms appear to constitute a spiral wave of higher than average density rotating through the disc of gas at an angular velocity slightly higher than the disc itself. This spiral instability in the gravitational field of the Galaxy seems to be driven by the motion of the stars, but a variety of completely different explanations have been suggested. That they represent outflow of gas from the galactic nucleus, for example, or that they represent tubes of enhanced magnetic field intensity. Recent work in the infrared and in molecular line astronomy has begun to reveal the dense, cool clouds of molecules (mainly hydrogen) and dust which are the potential sites of star formation. The passage of the spiral density wave appears to trigger the collapse of such a cloud and its fragmentation into a loose association or cluster of stars. But the details remain a mystery for the moment. We still do not know whether the heavy elements (carbon upwards) in these clouds are mainly in the form of molecules or are condensed onto dust grains, and the nature and role of the dust grains is still a matter for speculation. This ignorance of the details of star formation embraces also the formation of planetary systems — how common are these? — and the origin of that most significant grain of dust (for us at least), the earth.

Once formed, a star like the sun evolves by transforming hydrogen to helium through thermonuclear reactions and, in the later stages of its life, helium to carbon, nitrogen, oxygen and so on. However, even the respectable edifice of stellar evolution theory has begun to look rather shaky, with the failure to detect the expected flux of neutrinos emitted in certain nuclear reactions in the core of the sun. The final stages of a star's life are still obscure. Theory tells us that only stars of mass less than about 1.4 times that of the sun can find a stable, cold, final state as a white dwarf or neutron star (degenerate, compact objects which simply cool off indefinitely). The identification of pulsars (pulsating radio sources with periods ranging from milliseconds to several seconds) as rotating neutron stars was a triumph for theoreticians who had predicted the existence of neutron stars more than 30 years previously. Do more massive stars manage to shed their "excess" mass in more or less violent outbursts (novae, supernovae), or must they inevitably collapse inside their "horizon" to give a black hole? The X-ray source Cyg X-1 has aroused tremendous excitement due to its possible explanation as a black hole orbiting another, visible star. Certainly there appears to be a dark, massive, compact companion onto which gas from the visible star is falling and getting heated up to produce the X-rays. Pulsars and X-ray sources in binary star systems are examples of totally unexpected discoveries from the new astronomies of radio- and X-rays, which turn out to clarify our ideas about how stars evolve. The newest astronomies of infrared and of γ-rays will doubtless have further surprises in store.

The first discovery of the oldest of the new astronomies, radio-astronomy, was that some galaxies are the sources of powerful non-thermal radio emission. Despite decades of intensive observational and theoretical work, there is no agreed model for the origin of the energy in the nucleus of the galaxy or for the generation of the double sources often symmetrically placed at distances of up to 10^6 lt.yr from the galaxy.

This problem of violent events in galaxies was exacerbated by the discovery of quasars, radio sources identified with quasi-stellar objects with large redshifts, often variable both at radio and optical frequencies. Their redshifts are up to 10 times those found in the faintest galaxies, so that their optical luminosities have to be 100 times greater, if the Hubble law (distance proportional to redshift) is satisfied. This led to doubts about the cosmological nature of the redshifts, still not resolved. Associations with galaxies of much lower redshift suggested that the main contribution to quasar redshifts might be some intrinsic effect in the quasar. However, the consensus (for what that is worth) is now that the redshifts *are* cosmological, and that quasars are simply a more violent form of the activity seen in Seyfert galaxies. The latter also have an active, quasi-stellar core, but one that does not outshine the whole galaxy, as quasar optical cores must do. Naturally we must keep looking for decisive evidence of the link between quasars and galaxies, and that quasar distances are as huge as their redshifts suggest.

We must be glad that we have some sort of overall scenario for the evolution of matter in the universe, but we must remain continually critical of it. To be a true scientist is only to have an inkling of the full extent of man's ignorance.

SUGGESTED FURTHER READING

D. W. Sciama, *Modern Cosmology*, Cambridge University Press, 1971.
M. Rowan-Robinson, *Cosmology*, Oxford University Press, 1977.
Frontiers in Astronomy, readings from *Scientific American*, W. H. Freeman, 1970.
G. B. Field, H. Arp and J. N. Bahcall, *The Redshift Controversy*, W. A. Benjamin, 1973.
G. R. Burbidge and E. M. Burbidge, *Quasi-stellar Objects*, W. H. Freeman, 1967.
Confrontation of Cosmological Theories with Observation, I.A.U. Symposium No. 63, ed. M. S. Longair, D. Reidel, 1975.

Fig. 1. The spiral galaxy, M51. The spiral arms are delineated by bright, newly formed stars. (Photograph from the Hale Observatories.)

Fig. 2. The rich cluster of galaxies in Hercules, containing both spirals and ellipticals. (Photograph from the Hale Observatories.)

Fig. 3. The nearby radio-galaxy, Cent A (NGC 5128). Note the dust lanes, unusual for an elliptical galaxy. The overall extent of the double radio source is more than a hundred times that of the optical galaxy. The nucleus of the galaxy is a source of X-rays. (Photograph from the Hale Observatories.)

Douglas Gough

Lecturer in Astronomy and Applied Mathematics at the University of
Cambridge and Fellow of Churchill College, Cambridge.

Research interests include solar oscillations, the theory of pulsating
stars, convection theory and thermal properties of the Moon.

THE SOLAR INTERIOR

A decade ago the Sun was thought to be well understood. Astrophysicists had been lulled into complacency by the knowledge that the theory of the structure and evolution of stars could be made to agree with the gross properties of the Sun, and on the whole could be made to reproduce the broad features of the observations of other stars. But the relevant solar data were few: just the luminosity and radius. Detailed features of surface heliography were generally thought to be minor perturbations to a simple theoretical model. This may indeed be true, but we are now aware that the theoretical model is not quite correct, and that understanding surface features might provide us with the tools necessary for probing the solar interior.

CONTRACTION TO THE MAIN SEQUENCE

Like all other stars, the Sun is believed to have condensed from a large diffuse gas cloud. Its composition was therefore that of the cloud; principally hydrogen and helium, with all other elements totalling just 1 or 2 per cent by mass. The early stages of stellar collapse are not well understood, but there is strong evidence that a star eventually settles down to almost hydrostatic balance, with pressure supporting it against its own gravitational attraction. Under such circumstances it can easily be shown that there is a balance between the total gravitational energy Ω, which is negative, and the kinetic energy T of the stellar material, such that

$$2T + \Omega = 0.$$

This is the virial theorem of Poincaré and Eddington and is valid provided no other forces, such as electromagnetic forces, play a significant role in the hydrostatic balance. If the gas of which the star is composed is not degenerate, in the quantum mechanical sense, T is basically the energy in the random thermal motion of the gas particles and so provides a measure of the mean temperature.

The total energy of the star is

$$E = T + \Omega$$

which, by the virial theorem, is simply $-T$. As energy is radiated from the surface of a star, E decreases and consequently T rises. This is perhaps the most important property of a self-gravitating body, and leads to a natural accentuation of the temperature difference between the body and its surroundings. The star slowly contracts; compression raises the temperature of the gas, but only half the energy gained by the gas from the gravitational field is lost to outer space. The central regions of the star are compressed the most, under the weight of its outer envelope, and become hotter than the material near the surface. The evolution is controlled by the rate at which heat can be transported down the temperature gradient and radiated away at the surface.

This slow, so-called Kelvin–Helmholtz contraction might be halted by the gas becoming degenerate or by thermonuclear reactions replacing gravity as a source of energy to maintain the pressure in the stellar interior. In the former case pressure becomes almost independent of temperature. The star cools and shines less and less brightly without significant change in shape or size; it eventually goes out, and becomes like a planet. More massive stars liberate more gravitational energy per unit mass for the same relative contraction, and consequently become hotter. Nuclear reactions set in before the gas becomes sufficiently compressed to be degenerate, and the star adjusts its structure so that the thermonuclear energy generated in the core just balances the energy radiated at its surface. Such a star is said to be on the main sequence, and is in a phase of its evolution that is generally believed to be the simplest to understand. The Sun is in its main sequence phase now.

SOME THINGS THAT ARE KNOWN ABOUT THE SUN

The mass M (2.0×10^{33} g) and radius R (7.0×10^{10} cm) of the Sun can be inferred from the orbits of planets and the apparent size of the solar disc in the sky. From a knowledge of the solar constant and the distance between the Earth and the Sun, the radiant energy output or luminosity L (3.8×10^{33} erg sec^{-1}) can be deduced. Similarly the neutrino output L_ν from the nuclear reactions has been inferred, assuming all but a negligible fraction of the neutrinos incident on the earth originate in the Sun. The Sun's surface radiates almost as a black body, so its effective temperature of $5570°$ K, which is determined from its area and L according to Stefan's radiation law, is typical of matter temperatures in the photospheric layers from which most of the radiation is emitted. The relative abundances of most of the elements in the solar atmosphere have been obtained by analysing spectral lines in the emitted radiation, but the atmosphere is too cool for the absolute abundances X and Y of the most common elements, hydrogen and helium, to be determined accurately.

The solar surface rotates with a period of about a month. Other characteristic times that could be relevant to understanding the main sequence evolution can be deduced from the data listed above. First, a dynamical timescale, which may be characterized by $2\pi\sqrt{(R^3/GM)}$, where G is the constant of gravitation; it is simply the orbital period of a body near the solar surface. Its value is 2 hr 47 min and is of the same order, though somewhat longer than the free-fall time from the surface to the centre of the Sun. From the balance implied by the virial theorem it can be deduced that it is also of the same order as the sound travel time. Any deviation from hydrostatic balance would lead to a change in the dimensions of the Sun on this timescale. Another important time scale is that for heat to be transported from centre to surface. At the present time this is not very different from the characteristic Kelvin–Helmholtz contraction time E/L, because the Sun has not changed its structure significantly on the main sequence. Estimating E to be $0.3\, GM^2/R$, the value for a homogeneous sphere, yields a time of about 10^7 yr. The Sun is not actually homogeneous, but this estimate is of about the right magnitude. The time taken to convert the entire hydrogen content of the Sun into helium at the present luminosity can likewise be estimated from the above data and the binding energy of helium. It is about 10^{11} yr and estimates an upper bound to the total time the Sun can remain on the main sequence. The age of the Sun is thought to be about 4.7×10^9 yr. Since this is much greater than the Kelvin–Helmholtz time it is clear that the Sun has spent almost its entire life on the main sequence.

MAIN SEQUENCE EVOLUTION OF THE SUN

Conditions in the solar interior are inferred by constructing theoretical models. Simplifying approximations must of course be made. Because the Sun is seen not to change its dimensions appreciably on the dynamical time scale, and because the same is true too of other stars similar to the Sun but apparently younger, it is assumed that hydrostatic balance is always maintained. Rotation and magnetic fields are normally ignored, which is a good approximation if their surface values typify the values in the interior. This implies that the Sun is spherically symmetrical, which is a state of minimum energy, and is verified by observation to be true at the surface to about one part in 10^5. It is also normally assumed that the Sun is not contaminated by infalling interstellar matter.

From an assumed structure at the beginning of the main sequence, a model representing a star with the Sun's mass is considered to evolve for a 4.7×10^9 yr. Uncertainties in the theory occur in the opacity, which determines the rate at which radiation diffuses, in the nuclear reaction rates and in the degree to which the equation of state deviates from the perfect gas law. Just as significant are the uncertainties in the initial conditions. Because the thermal diffusion time is only a small proportion of the solar age, thermal balance is achieved early in the main sequence lifetime, so it is not necessary to know accurately the initial thermal structure. But it is important to know the initial composition. There is evidence that whilst contracting onto the main sequence the Sun experienced a phase of turbulent convection throughout its entire volume. Although this has been questioned it is normal to adopt an initial main sequence model that is chemically homogeneous, with a composition characterized by the constant abundances X_0, Y_0 of hydrogen and helium, the remaining elements having relative abundances equal to their current surface values.

During evolution hydrogen is converted into helium in the central regions. The models are found to be stable to convection in all but the outer 20 per cent by radius, so it is assumed that the products of the nuclear reactions remain *in situ*. Thus a helium-rich core develops. The number of particles per unit mass decreases, with a consequent reduction in pressure, which allows further slight gravitation contraction, an increase in density and temperature and thus an increase in the nuclear reaction rates, which are sensitive especially to temperature. This takes place on the nuclear time scale, much more slowly than thermal diffusion, and so the increased thermonuclear energy generation is reflected by an equal increase in the luminosity. Typical models predict that the luminosity has increased on the main sequence from about 70 per cent of its present value.

The convection zone in the outer envelope of the Sun presents a severe difficulty, because understanding of convection even in laboratory conditions is inadequate. A simple phenomenological theory is used with adjustable parameters that can tune the outer layers of the solar model.

The object of the calculation is to reproduce the relevant observations of the Sun. Until the end of the 1960s, these were just the present luminosity L and the radius R. With X_0, Y_0 and the parameters of the convection theory uncertain, it was possible to find many acceptable models. Indeed for any plausible abundance $Z = 1 - X_0 - Y_0$ of heavy elements, a value of X_0, can always be found that reproduces L and R once the convection theory is appropriately adjusted. Furthermore, the value of X_0 so obtained is in the vicinity of 0.7, which is similar to the hydrogen abundance of the solar wind and to the abundance deduced from analysing H and He spectral lines formed in the atmospheres of hot stars.

In the last decade two new observations have cast serious doubt on the validity of the simple models described above. The first measured the neutrino luminosity, and the second the oblateness of the solar image.

THE SOLAR NEUTRINO PROBLEM

In 1968 R. Davis and his collaborators dealt a severe blow to the theory. They attempted to detect neutrinos produced by certain reactions in the proton-proton chain, the chain of nuclear reactions believed to operate in the solar core. They were unable to detect any, but from a knowledge of the sensitivity of their apparatus were able to put an upper bound on the neutrino luminosity L_ν of the Sun which was a factor 10 below contemporary theoretical predictions.

The experiment depended on neutrino capture by ^{37}Cl stored in a 400,000-litre tank of perchloroethylene; the product is ^{37}Ar which was extracted chemically and subsequently detected as it suffered beta decay. It has become customary to measure L_ν in units depending on this experiment: a solar neutrino unit, or snu, is 10^{-36} captures per ^{37}Cl atom per second. Later refinements of experimental technique appear now to have led to a definite value for L_ν. Taking some four years of measurements into account yields 1.2 ± 0.5 snu, though measurements of order 4 snu have been obtained using the latest most sensitive detectors.

These results led theorists to check their calculations. The microphysics was thoroughly revised, and uncertain parameters were pushed to the limits of plausibility in order to produce as low a value of L_ν as possible. Now values just slightly in excess of 4 snu can be obtained with the so-called standard theory, which might be considered tolerable if all but the very latest measurements are ignored. If, however, the average of 1.2 snu is believed, the problem still remains.

In attempting to account for the discrepancy astrophysicists have questioned in turn all the assumptions of the theory. Of course, the most obvious starting-point is the nuclear physics. Since the results of the neutrino experiment were first announced all but one of the nuclear reactions in the proton-proton chain have been recalibrated. The remaining one is the first reaction of the chain. It proceeds via a weak interaction and is the slowest reaction of all, thus controlling the operation of the entire chain. The rate is too small to measure, and must therefore be calculated theoretically. Although many nuclear physicists seem to have confidence in the calculation, there is some experimental evidence from a different though similar reaction that the calculation contains a flaw. If that is the case, the neutrino problem disappears, but other investigations attempting to solve the problem have raised new questions about the Sun that now remain to be answered.

One such inquiry concerns the initial chemical composition of the Sun. It has been proposed that some fractionation process separated the elements as the Sun contracted from its parent gas cloud, and that subsequently no homogenizing fully convective phase took place to destroy the birthmark. Solar models with an initial inhomogeneity tailored to reduce the neutrino flux all have the property that in some region the density increases upwards. They are therefore subject to the Rayleigh-Taylor instability that occurs when one tries to create an interface between two fluids with the denser fluid on top. This deficiency might have been avoided had the composition gradients been smeared out somewhat. Because the material nearer the centre of the Sun is more highly compressed, it is possible to imagine a distribution in composition such that the mean molecular weight increases upwards without the density doing so. It would then be Rayleigh-Taylor stable, but it would be subject to the diffusively controlled fingering instability that occurs, for example, in the terrestrial oceans where hot salty water rests on colder, denser fresh water. Because chemical diffusion is slower than heat transfer, fingers of fluid thin enough to destroy their temperature differences but not so thin as to lose their chemical identity can penetrate vertically, liberating gravitational potential energy enough to drive the motion further. The process has been studied in the laboratory, but unfortunately it is not yet well enough understood to make reliable extrapolation to stellar conditions possible. Therefore it is not certain whether fingers would have had time to homogenize the Sun.

Another hypothesis is that substantial accretion of matter has taken place during the main sequence history of the Sun. Accretion of gas is unlikely, because it would be prevented by the solar wind, but interstellar gas clouds are known to contain dust which could fall through the wind. An obvious consequence of this hypothesis is that the resultant contamination of the solar surface, rich in heavy elements, would obscure the interior composition. Furthermore, solar models with interior heavy element abundances lower than the observational limits on the surface composition yield neutrino fluxes as low as the bounds set by Davis. Of course, this situation too would be subject to the fingering instability. It has been suggested also that the gravitational energy released during the intervals of accretion would temporarily increase the solar luminosity and upset the Earth's climatic balance, inducing glaciation. Dust clouds exist in the spiral arms of the galaxy, and the interval between successive passages of the Sun through the spiral arms is of order 10^8 yr, which is similar to the period between the major terrestrial ice ages. Moreover, the fact that the Sun is currently at the edge of a spiral arm is consistent with the Pleistocene glaciation of the last 10^6 yr. These coincidences are intriguing, but it is difficult to reconcile the large amount of accretion required by the theory with the estimates of the amount of dust in the spiral arms.

A third investigation concerns the stability of solar core. Although the core is stable to convection, regions in which thermonuclear energy is being generated are potentially unstable to oscillatory gravity waves or acoustic waves. Each element of fluid can be thought of as a thermodynamic engine, generating heat preferentially at high temperatures, and thus doing work that drives the oscillations to greater and greater amplitude. Calculations suggest that only after equilibrium concentrations of the intermediate products of the proton-proton chain have been built up over a sufficiently large volume of the core can the instability occur. This takes about 10^8 yr. It is then presumed that the oscillations break down into turbulence and mix material rich in fuel from the edge of the core into the centre. This temporarily enhances the energy generation rates and upsets the balance of the reaction chain in such a way as to quench the instability. The Sun is then quiescent again until nuclear equilibrium is once more achieved, and the whole process repeats.

An immediate consequence of the enhanced energy generation in the core is an expansion of the star. It is the reverse of the gravitational contraction that occurs when nuclear energy sources are absent or inadequate. The expanding sun cools and the luminosity decreases by a few per cent. The surface luminosity differs from the nuclear energy generation rate until thermal balance is restored on the Kelvin–Helmholtz time scale of about 10^7 yr. During this time the solar core, temporarily cooled by its expansion, liberates fewer neutrinos than usual. Once again, if the luminosity fluctuations are presumed to induce ice ages on the Earth, the beginning of the Pleistocene epoch marks the most recent mixing of the Sun's core. Because that occurred less than 10^7 yr ago, it is inferred that the Sun is presently in one of its comparatively rare states of thermal imbalance, and the neutrino flux is anomalously low.

Further work is necessary before these ideas can be reliably tested. The ^{37}Cl neutrino experiment detects mainly high-energy neutrinos produced in a side reaction of the proton-proton chain, and does not provide a direct measure of the heat generated. An experiment using a gallium detector has been proposed that will measure the lower energy neutrinos from the thermal energy-producing chain. This could provide a new check on the nuclear physics, and might enable one to decide whether or not the solar luminosity is currently balanced precisely by thermonuclear sources.

THE SOLAR OBLATENESS

The surface of the Sun is observed to rotate. This is inferred from the motion of sunspots and from line-of-sight surface velocities measured from the Doppler shifts of spectral lines. The rotation is not rigid: the equatorial regions rotate with a period of about 25 days whereas near the poles the period is of order 35 days or more. Moreover, sunspots and some other magnetically related phenomena that are presumably anchored by magnetic fields to material beneath the surface rotate somewhat more rapidly than Doppler measurements indicate, at least in the equatorial regions. This suggests an increase in rotation rate with depth. Nevertheless, it is generally believed that the rotation in the deep interior is not a great deal faster than it is at the surface.

An important property of rotating bodies is that centrifugal forces cause them to bulge at the equator. The effect on the Sun is clearly not very great, because the Sun appears circular to the eye. Indeed, if rotation were uniform throughout the Sun with a period of 25 days the polar and equatorial radii would differ by only about 1 part in 10^5. Consequently for most practical purposes the Sun can safely be assumed to be spherically symmetrical. This implies, in particular, that the solar gravitational field is also spherically symmetrical, and independent of how mass is distributed within the Sun. This important result has enabled astronomers to calculate with great precision the orbits of the planets without requiring detailed knowledge of the solar interior. However, it was realised as long ago as the middle of the last century that the orbit of Mercury was not what would be expected from Newton's inverse square law of gravitation. The discrepancy, after taking due account of perturbations from other planets, can be expressed as a precession of the perihelion of the orbit at a rate of 43 seconds of arc per century. Though small, this precession was a cause of anxiety to astronomers, and led them to predict the existence of another planet, Vulcan. The planet was never discovered. In 1915 Einstein predicted a precession of just the right value from the correction to the inverse square force law implied by his theory of general relativity, and the matter appeared to be settled.

In 1961 Einstein's explanation was challenged. C. Brans and R. H. Dicke proposed an alternative theory of gravitation which predicted a smaller precession rate for planetary orbits. It was suggested that only by chance had Einstein predicted the correct value, for the centre of the Sun is in a state of rapid rotation which distorts the gravitational field and so adds a contribution to Mercury's precession. To explain the observed precession by the Brans–Dicke theory the rotation period of the solar core would have to be about a day, and general relativity would then yield a precession rate too great by about 4 seconds per century. Of course, such rapid rotation would also distort the shape of the Sun's surface. To check this Dicke, together with H. M. Goldenberg, measured the shape of the solar image and, arguing that the edge of the sun coincides with a gravitational equipotential surface, found it to be oblate by just the amount required to explain Mercury's orbit with the new theory.

The experiment initiated much discussion. If the interpretation were correct, why should the solar core rotate so rapidly? A possible answer came from observations of other stars, younger though otherwise similar to the Sun. Although it is not possible to resolve their surfaces, Doppler broadening of their integrated light can be measured and rotation rates inferred statistically if it is assumed that the only variable in an apparently homogeneous class of stars is the angle of inclination of the rotation axis to the line of sight. What emerges from such an analysis is that surface rotation decreases with age. This is not a surprising result if one bears in mind that the Sun is continually ejecting a diffuse stream of gas, or wind, as are presumably other similar stars. There is a weak magnetic coupling which transfers angular momentum from the Sun to the wind and slows the Sun down. The present torque can be measured. Plausible estimates of how it has varied rationalizes the observed dependence of stellar rotation with age and suggests that the Sun might well have arrived on the main sequence rotating with a period of about a day. The current rotation period of the solar core might still be of that order if the coupling with the rest of the Sun is extremely weak. But is that so?

Dicke estimated how fast the core ought to slow down on the assumption that viscous shear stresses provide the only vehicle for angular momentum transfer beneath the outer convection zone. The characteristic time scale exceeded the age of the Universe. But the relevance of the calculation was disputed because a similar argument applied to a stirred cup of tea predicts a deceleration time as long as an hour. It is well known that tea stops rotating in about a minute. The explanation stems from an imbalance of pressure and centrifugal forces near the base of the teacup where viscous forces prevent rotation of the tea; this drives a circulation in meridional planes, originally discovered by V. W. Ekman, that advects angular momentum through the body of the fluid faster than is possible by viscous shear alone. It is also responsible for the convergence of the tea leaves in the middle of the cup. So perhaps the solar core is similarly decelerated and general relativity is correct. Interestingly, the dynamics of the Ekman circulation has been discussed by Einstein, in a paper on the meandering of rivers. It is not easy to extrapolate the analysis of cups of tea to the Sun, however, particularly as the Sun is not in a rigid container and is not of uniform density. None the less, it does seem likely that a similar circulation is set up, though how it redistributes angular momentum is less clear. Thus, even granted that the inside of the Sun feels the deceleration of the outside, theory cannot yet predict what the rotation of the inside ought to be.

One of the problems encountered in measuring the shape of the Sun's gravitational equipotentials by observing the shape of the solar image is in deciding just what one means by the edge of the Sun. The difficulties have been stressed recently by H. A. Hill and his collaborators, who observed that the spatial variation of the intensity of radiation near the edge of the Sun depends on latitude: near the equator it declines more sharply than it does at the poles. Hill interpreted this as resulting from an excess brightness just above the photosphere, rather than from variations in matter density, which could arise from the preferential dissipation of atmospheric waves above the photosphere in the equatorial regions. Taking this into account Hill inferred a value of the oblateness consistent with almost uniform rotation of the solar interior. But that is not all. In the course of his observations Hill discovered that the radius of the Sun was changing with time: the Sun was oscillating on a dynamical time scale. Higher-frequency oscillations had been well studied before, with periods corresponding to those of waves confined to the atmosphere alone, but these new oscillations have periods characteristic of motion whose amplitude penetrates deep into the Sun. Analysis of the data has yielded a spectrum of frequencies that agrees well with theoretical frequencies of normal modes of oscillation of a solar model computed with the standard assumptions listed earlier in this article.

The existence of these oscillations has an interesting implication regarding the solar rotation. It may be they that transmit energy above the photosphere and generate the equatorial excess brightness. Their latitude dependence has not yet been measured. However, if they are responsible for much of the apparent oblateness the oscillations must have greatest amplitudes near the equator. Such modes also have the property of being asymmetrical about the rotation axis, and of propagating around the Sun in longitude. In particular there are classes of such modes, whose members each have frequencies that are almost identical, but with slight differences induced by the Sun's rotation. Two such modes oscillating together would beat, and the dissipation would give the Sun the appearance of having equatorial bulges that rotate in longitude at a rate depending on the modes that are beating. Such a phenomenon has recently been noticed by Dicke, who observed that the apparent oblateness varies with a period of 12.2 days. At present the measurements are not detailed enough to enable the modes that are beating to be identified, or even to confirm or refute the hypothesis that it is indeed a beat phenomenon that is being observed, but if the hypothesis is correct it is clear that the rotation rate of at least some region of the solar interior exceeds the rotation of the surface, though not by a large amount. Hopefully future observations of this kind will make more precise deductions possible.

CONCLUDING REMARKS

The neutrino flux and oblateness measurements have shocked astrophysicists out of their comfortable belief that the Sun is a simple, easily understood body. None of the theoretical models of the time was able to explain the neutrino observations, which provided a stark reminder that the ability of a theory to rationalize data does not prove that theory. Even if it turns out that the neutrino luminosity is not as low as has been feared, and, as now seems likely, that the oblateness of the gravitational equipotentials is no greater than would be expected from assuming that the rotation of the solar surface typifies the rotation of the Sun as a whole, these observations have been of great importance because they have stimulated much deeper thought into the problems posed by trying to infer the structure of the solar interior. Furthermore, they have led to the discovery of a spectrum of coherent oscillations which may provide a powerful diagnostic tool for probing the Sun, just as seismic oscillations have given important information about the interior of the Earth. In particular, it may be possible to determine whether the Sun has been contaminated by accreted interstellar material.

Why the oscillations should even be present is not understood, though it is likely that they are triggered by turbulence in the convection zone. The rough agreement between the theoretical and observed frequencies suggests that current thinking is not altogether incorrect, but the differences do appear to be significant. Perhaps the discrepancies will be removed by minor changes in the theoretical models, but alternatively they may be linked with other phenomena, such as magnetic fields, hitherto ignored by most theorists. L. Mestel has pointed out that quite strong magnetic fields may reside deep in the interiors of some stars, even though the surface fields may be quite small. If sunspots could be understood, they might provide a clue to the interior field of the Sun. The cyclic behaviour of the Sun's surface magnetism, with a period of 22 years, is a solar phenomenon without a convincing explanation, and might eventually provide additional diagnostics.

The discussion of the solar oblateness has brought to the forefront an interesting problem: the history of the Sun's rotation. Even though the question of angular momentum loss by the Sun had previously been considered, it was not until after Dicke's oblateness measurement that mechanisms coupling the core to the surface were seriously investigated. Magnetic fields and the coherent solar oscillations are capable of angular momentum transport. So is rotationally induced material circulation. The possibility of the latter raises new questions concerning mixing of the products of nucleosynthesis into the solar envelope.

Perhaps the new solar data have generated more questions than answers. They have certainly reinforced our awareness of our ignorance of the solar interior.

FURTHER READING

Bahcall, J. N. and Davis, R. Jr. (1976). Solar neutrinos: a scientific puzzle. *Science*, **191**, 264.

Bahcall, J. N. and Sears, R. L. (1972). Solar neutrinos. *Ann. Rev. A. & Ap.* **10**, 25.

Bretherton, F. P. and Spiegel, E. A. (1968). The effect of the convection zone on solar spin-down. *Ap. J.* **153**, 277.

Christensen-Dalsgaard, J. and Gough, D. O. (1976). Towards a heliological inverse problem. *Nature*, **259**, 89.

Dicke, R. H. (1970). Internal rotation of the Sun. *Ann. Rev. A. & Ap.* **8**, 297.

Hill, H. A. and Stebbins, R. T. (1975). The intrinsic visual oblateness of the Sun. *Ap. J.* **200**, 471.

Tayler, R. J. (1970). *The Stars: their Structure and Evolution*, Wykeham, London.

Paul Charles William Davies

Lecturer in Applied Mathematics, King's College, University of London.

Research activities have covered several areas of fundamental physics including quantum theory, relativity and cosmology.

Recent research has been on quantum field theory in curved space, and particularly concerned with quantum particle production effects near black holes and during the initial moments after the cosmological big bang.

CURVED SPACE

For those who think of space as emptiness, the assignment of an adjective, especially one as enigmatic as "curved", might be regarded as cryptic. To the mathematician, and more so to the physicist, empty space may be devoid of matter, but it is by no means devoid of properties. I shall, at the outset, sidestep the dangerous mire dealing with the centuries-old controversy over whether space (and time) is really a substance existing in its own right, or simply a linguistic convention describing the relations between material objects. The question of the curvature of space may only be settled observationally by the use of light rays and material structures, and it matters little here whether the results obtained are to be regarded as properties of the latter or not.

What does "curved space" mean? Curvature is a familiar property of everyday life. But there is a subtlety; *lines* can be curved, and *surfaces* can be curved. Two quite distinct entities may share a common property. We describe with the same adjective (and scarcely a thought) things as fundamentally dissimilar as a railway line or the surface of the Earth. The distinction here is one of dimensionality. A point on the one-dimensional railway line needs but one number (e.g. the distance from Euston station) to locate it uniquely, while a point on the two-dimensional surface of the Earth needs two numbers (e.g. latitude and longitude) for its location. Here I shall explore the question: can curvature apply to *three-dimensional* volumes, as well as two-dimensional surfaces and one-dimensional lines, and if so, what can be learnt from observation and mathematical theory about it?

Although the examples of railway line and Earth describe the curvature of material entities, this is not an essential feature; for example, the path of an arrow is a set of points in space which possesses curvature. The problem of curved volumes is just the problem of curved space, empty or otherwise.

In mathematics, curvature belongs to the subject of geometry. School geometry is called Euclidean, after the Greek geometer Euclid. Everyone remembers a few of the theorems — the angles of a triangle add up to two right angles being an elementary one. Euclidean geometry works well on *flat* sheets of paper. It does not always work on *curved* surfaces, such as that of the sphere. A glance at Fig. 1 shows how triangles on a sphere may have three right angles!

Until the nineteenth century nobody questioned the fact that although, for example, the Earth's surface was curved, the surrounding space itself was subject to the rules of Euclidean

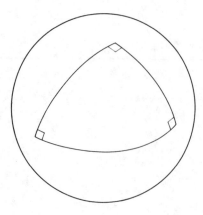

Fig. 1. Two-dimensional spherical space. The usual rules of Euclidean geometry do not work here. A triangle may have three right angles.

geometry. Indeed, ever since Kepler physicists had frequently employed the Euclidean theorems across the empty space of the solar system to describe the paths of the planets around the Sun. "Flat" space geometry seemed to work well in these regions remote from the Earth. With one exception. Mercury, the nearest planet to the Sun, seemed to have an orbit which was slightly distorted. This distortion could not be accounted for by the disturbance of the other planets, and although exceedingly small (a change of only 43 seconds of arc per century) remained a total mystery until the beginning of this century.

Already the nineteenth-century mathematicians Gauss, Lobatchevsky and Riemann had developed non-Euclidean geometry for volumes, but it took the genius of Einstein to construct a *physical* theory which exploited these mathematical developments. Einstein was interested in the motion of material bodies, and had constructed a theory of uniform relative motion, called special relativity, in 1906. In attempting to generalise relativity to accelerative motions, he realised that the unique nature of gravity singled it out from all other physical forces which accelerate material bodies. This is because (small) bodies which fall freely under gravity all travel along the same path in space, no matter what their mass or constitution. This suggested that gravity was not really a force at all, but a geometrical property of space itself (strictly speaking, space and time together). Einstein's revolutionary proposal was to identify gravity with a departure from Euclidean geometry — in popular parlance, a bending or curvature of space. According to this viewpoint the planets do not move in a curved path around the Sun, as Newton supposed, with the Sun exerting a gravitational *force* upon them to bend them away from their natural tendency to move in straight lines. Instead, the Sun's gravity is interpreted as a distortion of space (and time) in its vicinity, and the planets merely follow the "easiest" route — the path which minimises their mechanical action through the curved space. This easiest route turns out to be very nearly the same as the "forced" route taken by the planets according to Newton's theory of a gravitational force. But not quite. Mercury's orbit has to be displaced by 43 seconds of arc per century. This was Einstein's great triumph.

If the space around the Sun is not precisely Euclidean, it might be expected that the images of objects seen beyond the Sun would be somewhat distorted. In 1919 Sir Arthur Eddington checked this prediction during a solar eclipse, by observing a displacement in the positions of stars in the region of the sky near the eclipsed Sun.

For many years this remarkable theory of Einstein, called general relativity, appeared to be limited to describing minute corrections to Newtonian gravity. This was because our knowledge of astronomy did not suggest that there were ever situations in the Universe where gravity was strong enough to cause a really drastic bending of space. However, during the last 10 years, this prospect has become entirely plausible. We shall see that it leads to some of the greatest outstanding mysteries of physical science.

Although the present density of gravitating matter in the Universe is very low (about one star of solar mass per billion cubic light years) the Universe is very big (present telescopes see as far as a billion light years) and the effect of space curvature can be cumulative, with each star causing its own tiny distortion. As long ago as 1915, Einstein realised that on a cosmological scale the cumulative space curvature might become so great that it would alter the *topology* of space. To understand this remark, return to the two-dimensional analogy. If the curvature of a surface (or sheet) is always in the same direction, and everywhere roughly the same amount, the surface will eventually join up with itself. The sphere is a good example of this. Although in a small enough region the geometrical properties of a spherical surface are not greatly different from those of a flat sheet, the *global* structure is distinctly different. For one thing, the spherical surface is *finite* in size, though nowhere does it possess a boundary or barrier. One

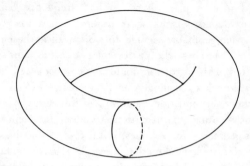

Fig. 2. Two-dimensional toroidal (doughnut) space. Unlike spherical space there are closed curves which cannot be continuously shrunk to a point.

consequence of this is that if we were to travel along the shortest route between any two points on such a surface, and then continue "straight" on, we should eventually return to our starting-point from the opposite direction.

The sphere is by no means the only finite and unbounded curved surface. The torus (see Fig. 2) is another example. This is distinguished from the sphere by the property that there exist closed lines on the torus which cannot be shrunk continuously to a point. Still more complex properties could be achieved by putting twists into the surface before joining them up, so that a journey round certain closed paths would change right-handed orientations to left-handed orientations.

Einstein's novel proposal was that space curvature closes up the Universe into the three-dimensional analogue of the spherical surface. Such a universe would then be finite in volume, and could be circumnavigated by a sufficiently tenacious traveller. It might even be possible to see the back of one's head by looking out into space through a large enough telescope! Still more bizarre would be a "hypertoroidal" space (like a three-dimensional doughnut) or even a "twisted" space, in which our adventurous traveller would return with his left and right hands transposed!

In spite of the Alice-in-Wonderland quality of these considerations, they are by no means idle curiosities. A determination of the global nature of space is one of the foremost tasks of modern cosmology, and a variety of observational techniques are being employed on this task.

The portion of the Universe accessible to even our largest telescopes is too small to reveal the global structure directly, so a combination of theory and observation has to be used. Since Einstein's original proposal, it has been discovered that the Universe is expanding — growing

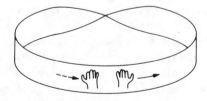

Fig. 3. Two-dimensional twisted space (called a Möbius strip). If both surfaces of the strip are identified as the same space, then a journey once round the loop changes right handedness into left handedness.

with time. If it is also spatially closed, space would then be analogous to the surface of a balloon being slowly inflated. It turns out that the *motion* of the Universe is related to the geometry of space through Einstein's theory of general relativity. Thus, it may be possible to determine if space is closed and finite by observing how the Universe moves.

This is no mean task. Distances between the galaxies only grow at about one-millionth of 1 per cent each century. However, because of the enormous size of the observable Universe, the light received from the farmost galaxies started out many millions of years ago, and it is possible to infer how the expansion rate then compares with now. Until recently, all indications were that the rate of expansion was gradually slowing down, and would eventually stop. This not only implied that space was indeed closed up on itself as Einstein had suggested, but also that the Universe would eventually collapse as recontraction followed the cessation of expansion. Recently, these observations have been questioned on a variety of grounds, with the result that the global geometry of space and the eventual fate of the Universe are still in the balance. What is needed is either a very good estimate of the amount of gravitating matter in the Universe, and this includes much which is inconspicuous (dark gas and dust for example), or else a better knowledge of the way galaxies change their brightness with age. The latter enables a better estimate of galactic distances, and therefore motions, to be made. The next decade or so should see this question answered.

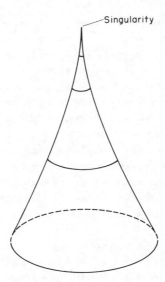

Fig. 4. A singularity in space. The circles on the cone-like structure measure the curvature of this two-dimensional space. As the apex is approached, these circles become progressively more strongly curved, until they reach a point of infinite curvature at the tip, where the space ceases to exist.

In 1967 J. Bell and A. Hewish of Cambridge University discovered a new type of star, called a pulsar on account of the regular radio pulses which they emit. It is widely believed that pulsars are incredibly compacted stars, so dense that they are only a few kilometres across and can spin many times a second. The compaction of matter in these stars is great enough for their atoms to be entirely crushed into neutrons. From what is known about nuclear matter, these neutron stars appear to be teetering on the edge of disaster. So intense is the gravity at their surface, that any incipient neutron stars more massive than our sun would not reach a stable structure at all, but would instead implode catastrophically, and in a fraction of a second disappear entirely from the Universe!

The explanation for this astonishing phenomenon lies in the violent curvature of space which accompanies the escalating gravity near a collapsing star. As the radius of the star shrinks, the space curvature at the surface soon becomes strong enough to bend back even light rays — the swiftest of all entities. After the last light ray has escaped the star becomes a prison — nothing whatever can escape to the outside, for nothing can travel faster than light. From a distance, the star appears to have been replaced by a hole in space — a black hole. Inside the black hole, the inexorable shrinkage continues; in fact, no force in the Universe can prevent the collapsing star from crushing itself ever more forcefully. With each microsecond, the space becomes progressively more curved and the density of the star rises.

The burning question is: what happens in the end? Does the implosion stop, or does the star literally shrink to nothing?

According to the general theory of relativity, the star must continue to shrink until space becomes so curved that a *singularity* forms. This situation may be visualised using the two-dimensional analogy of a surface which has been curved to form a cone-like structure. The curvature of the surface rises without limit as we pass towards the point of the "cone". At the point itself something catastrophic occurs — the curvature is, loosely speaking, infinite, and the "space" comes to an end.

Can it really be true that space inside a black hole is so curved that it too comes to an end? This is not just an idle question. Even if our present understanding of stars is badly wrong, a black hole could in principle be made by a sufficiently resourceful and determined technological community. Moreover, the present picture of an expanding universe leads almost inevitably to the conclusion that this expansion began in a big bang some 10 to 20 billion years ago. The condition of the Universe then was very similar to the inside of a black hole, though in reverse. The Universe apparently emerged from a singularity before which neither space nor time existed. This scientific picture of the creation of the Universe cuts right across philosophy and religion, and is surely today's most outstanding challenge in physical science.

Whether or not these far-reaching conclusions about space and time terminating are believed depends on the extent to which Einstein's theory of relativity can be trusted to apply as successfully over very small distances as it evidently does over cosmic dimensions. This is the question which theoretical physicists would dearly like to answer. Let us see how well it has already been answered.

One way in which a black hole singularity might be avoided is if the implosion is not symmetric. Then different parts of the star would fall in slightly different directions, thereby "missing" each other at the centre. Alternatively, some drastic modification might occur in the properties of the contracting matter (which is, after all, compacted way beyond all terrestrial experience). Remarkably enough, neither of these two possibilities proves particularly fruitful. Some amazing mathematical theorems, largely due to the British mathematicians S. Hawking and R. Penrose, demonstrate that, under a very wide range of circumstances, a singularity will inevitably form. The sort of modification to the properties of matter necessary for singularity avoidance are most unacceptable to physicists, involving negative energy matter and so forth.

It might then be asked: will Einstein's theory break down in the extreme conditions which precede the singularity? We cannot give the definitive answer "no" without falling into one and being crushed out of existence. But the answer "yes" could be near at hand. There is a sense in which the general theory of relativity is known to be wrong as it stands. This is because it is incompatible with the laws of microscopic physics, known as *quantum mechanics*. The quantum theory has been responsible for averting another catastrophic collapse – the implosion of atoms as a result of electric attraction. According to pre-quantum ideas of mechanics, all atoms should be unstable and collapse, but they obviously don't. The resolution of the paradox is provided by the quantum theory, which explains the stability of atoms by supposing (heuristically) that microscopic particles are subject to sudden fluctuations in their energy and position. These fluctuations have the effect of preventing the particles in an atom from approaching each other too closely. So, could this effect also operate in gravitational collapse, when the star has shrunk down to atomic dimensions?

The answer is, not directly. The quantum properties of matter alone are probably not strong enough to overcome gravity. But there is still the possibility that the quantum properties of *gravity itself* might be important. Fluctuations in gravity imply fluctuations in the curvature of space. If such microscopic fluctuations really occur, they could profoundly modify the structure of space on a very small scale, for the smaller the distances of interest, the larger the fluctuations become. Eventually, they would be so violent that space would literally start to tear itself apart. The sort of dimensions on which such a traumatic effect would be operative is minute indeed – a hundred billion billion times smaller than the nucleus of an atom! At this scale, space would no longer have uniform properties but would assume a sponge-like structure – with bridges and holes interconnecting in a fantastic and complex web. Just what happens to a catastrophically collapsing body which is squeezed into this spongy domain is anybody's guess. Whether space and all of physics ends at a quantum singularity, or whether space gives way to some new entity, and if so what, are questions eagerly under attack by theoretical physicists.

During the last 20 years, mammoth efforts have been made to formulate a rigorous theory of quantum gravity to give a proper theoretical basis to these conjectures. Unfortunately, not only does success seem very far away, but some physicists even doubt whether the quantum theory is compatible at all with the general theory of relativity. Such an incompatibility would presage a new theory more fundamental than either of the old, involving unfamiliar concepts out of which curved space must be built as a "large"-scale approximation. To find out just what these concepts might be, and what strange new perspectives they will provide on the nature of the Universe, we must await tomorrow's discoveries.

I. W. Roxburgh

Professor of Applied Mathematics at Queen Mary College, University of London, and Chairman of the London University Astronomy Committee.

Member of the Royal Astronomical Society and the International Astronomical Union, Member of the Council of the Royal Astronomical Society 1968-71.

Research interests include the History and Philosophy of Science, Cosmology, Evolution of Stars, Solar Physics and Gravitational Theory.

IS SPACE CURVED?

Ever since the development of general relativity gravitation has been attributed to the "curvature" of space, in the words of John Wheeler: "There is nothing in the world except empty curved space. Matter, charge, electromagnetism are all manifestations of the bending of space. Physics is Geometry."[1] While this is a rather extreme example of the geometrodynamic philosophy it is nevertheless shared by the scientific community, and the belief that "space is curved" has entered and remained in our conceptual framework of physics. But is space curved? Indeed what if any meaning can be given to such a statement? I shall argue that such a statement is meaningless, empty of any empirical content, space is curved or flat depending on the choice of the scientist.

Until the discovery of non-Euclidean geometry in the eighteenth and nineteenth centuries the geometry of space was assumed to be the flat Euclidean geometry discovered by the Babylonians, Egyptians and Greeks, finally formalised by Euclid in the *Elements*. Drawing on practical experience of measurement and surveying they discovered that the sum of the angles of a triangle was always 180°, that the square of the hypotenuse of a right-angled triangle was always equal to the sum of the squares of the other two sides, that the circumference of a circle was always the same multiple of the diameter. These results were acquired through measurement, using a ruler or angle measuring device — it was an empirical geometry. From Euclid onwards the situation was changed, by showing that all the empirical results of geometry could be deduced logically from a small number of definitions and axioms, it set the example for axiomatic mathematics and demonstrated that the behaviour of the physical world was governed by a few simple statements, or laws.

Then came the discovery of non-Euclidean geometry — not as the result of measurement but as the consequence of logico-mathematical inquiry, by changing one of the axioms in the Euclidean scheme an alternative astral, or hyperbolic geometry was discovered, just as logically sound as the Euclidean example. We see geometry now as simply a mathematical structure, given certain definitions and certain axioms — then the application of standard logical reasoning produces a geometry. There is no end to this procedure — any number of different mathematical geometries can be (and have been) derived by mathematicians.

But is there one of them that corresponds to the real world — a real geometry of space? Shortly after his discovery of hyperbolic geometry, Karl Fredrick Gauss set out to try and answer this by direct surveying techniques, from points on the mountains of the Brocken, the Höher-Hagen and the Insleberg, he measured the sum of the angles of a triangle to see if they were equal to or less than 180° and space was flat. Gauss like others since have been convinced that the determination of the curvature of space is an empirical issue, there is this substance called space that has intrinsic characteristics — one of which is curvature — and the experimental scientist can in principle measure that curvature.

This belief that space has substance of its own has a very long history, but was finally accepted into physics with the success of Newtonian physics and the Newtonian philosophy of nature. But it was challenged in Newton's own time by Gottfried Leibniz who advocated a relational theory of space and time;[2] only the relationships between objects or time ordering has any empirical meaning — to introduce a concept of absolute space with no properties other than that of being absolute he considered a redundancy, indeed a challenge to the absolute power of God. But in spite of the cogency of Leibniz's arguments, Newton's natural philosophy prevailed — because it worked — no alternative to Newton's physics was available and scientists accepted the philosophy as well as the physics of the Newtonian scheme.†

---

† This was a problem Huygens wrestled with for many years but he was unable to account for Newton's absolute centrifugal forces.

Criticism of the Newtonian philosophy arose again in the writings of Ernst Mach that so influenced Einstein in his search for general relativity, absolute space was meaningless, inertia in matter must mean inertia relative to other matter not relative to absolute space.[3] But absolute space remained in physics through the revolution of Special relativity in which absolute space remained as the backcloth defining the inertial frames in which the new Lorentz invariant physics was valid. While special relativity has a major impact on physics, conceptually it is little different from the old Galilean Newtonian physics, replacing Galilean invariance with respect to inertial frames by Lorentz invariance but still retaining the absoluteness of space and inertial frames.

With the advent of general relativity Einstein hoped to give expression to the views of Leibniz and Mach in a quantitative way, to show how the inertial properties of matter were indeed caused by the rest of the matter in the Universe, absolute space was to be abolished in favour of relative space, there was to be no absolute space — no privileged reference frame — anyone's view was as good as anyone else's and only relative behaviour entered physics. Unfortunately this programme did not succeed, absolute space remains in general relativity, but even more, general relativity described the inertial gravitational field in terms of the curved space-time geometry developed by Riemann following on the work of Bolyai, Lobatewski, and Gauss.[4] This was an ideal tool for the expression of these views, the inertial and gravitational properties could be described in terms of a curved space, the curvature produced by matter, and bodies moving along "straight lines" in this geometry. The theory, like Newton's before it, works, that is its predictions agree with experiments, and just as in the Newtonian period scientists have accepted not only the theory, but the philosophical backcloth of curved space-time, just as they accepted Newton's absolute space. We live in a curved space — the scientist's task is to determine the curvature.

But what is this stuff called space whose curvature is to be determined — how do we measure it? We can, like Gauss, set up a triangulation experiment and measure the angles of a triangle — the answer will not be $180°$ — but this does not mean space is curved. The experiment is done with light rays and theodolites — the empirical result is a statement about the behaviour of light rays — not about space. It is, as it must be, an experiment about the relationship between objects in space not about space itself. The same is necessarily true about any experiment; from it we learn of the relationship between objects not of the background we call space. Indeed it is part of the philosophy of relativity that it is the relative positions and motions of matter that determine the behaviour of other matter; space, or space-time, is an intermediary that we bring into the formalism for ease of representation, but in any empirical statement about the world the representation is eliminated, in the end scientific theory is reduced to numerical predictions of the kind $A = 2B$, where A is the predicted quantity and B a reference quantity. But the method of representation has been given an existence in its own right — we talk of curved space, or curved space-time, as though this is something intrinsic to the world instead of the mathematical representation it actually is.

The difference between the mathematical representation and empirical physics was carefully elucidated by Henri Poincaré[5] at the beginning of this century, and led him to his conventionalist philosophical stance. Clearly distinguishing between experimental data, and the mathematical framework used for expressing them, Poincaré announced his dictum "No geometry is more correct than any other — only more convenient".[5] To illustrate this point consider an accurate version of Gauss' surveying experiment, the sum of the angles of the triangle is greater than $180°$, as predicted by general relativity and in agreement with curved non-Euclidean geometry. How are we to interpret this result? The relativist announces that space is curved, but his actual experiment is with light rays and carefully stated is that the sum of the angles between the light rays is, say, 180.1°, it makes no mention of space. The experiment can equally be interpreted by saying that space is flat and Euclidean but that light rays do not move on straight lines. To build up a description of the world the scientist will then look for objects that influence the motion of photons — he will find them — the mass of the Earth, the Sun, the Moon, etc., he can now construct an equal representation of the physical world taking space to be flat and Euclidean. No experiment can decide between his representation and that of the relativist for they make the same predictions about empirical consequences but they choose to express it in different languages, for one space is flat, for the other it is curved — yet there is no way of choosing between them. The curvature of space is therefore in the mind of the scientist, to be chosen at will in the description of the physical world. The geometry or mathematical framework is like a language — an empirical truth about the world can be expressed in many languages, the truth is independent of the means of expressing the truth. The physical world is no more German because Einstein expressed his theory in German, than it is curved because he expressed it in a curved space-time.

We can build on this analogy by drawing up a dictionary to translate from the curved space-time description to the Euclidean description.

Curved space	Euclidean space
Space curvature	Fields in flat space
Straight line in curved space	Curved line in flat space
Matter causes curvature	Matter causes fields
Field equations for curvature	Field equations for fields

It follows, therefore, that not only can physics be described in flat Euclidean space but indeed in any space that the scientist chooses. It may be that the theory can be expressed more succinctly in one geometry, but that does not make that geometry "correct", only convenient. The curvature of space is at the behest of the scientist.

Nowhere has the "curvature" of space had a firmer hold than in cosmology, the large scale structure of the Universe. General relativity predicts three possibilities, a positive curved space, negatively curved, or flat, and experimental cosmologists strive to determine this curvature. But again, this curvature is an illusion. A given experiment yields a particular answer about the properties of objects in the Universe — it cannot tell us anything about the curvature of the Universe, only about the results of experiments and we can choose how to interpret these. This was realised many years ago by E. A. Milne[6] who showed that his particular model of the Universe could be described either as a static infinite negatively curved Lobatewski space, or as an expanding motion of galaxies in a flat Euclidean space, depending on the units of measurement, both representations are equally valid and make the same predictions about any experiment, but by changing the units of measurement, that is our conventions, we change the curvature, only if we accept one set out of an infinity of possible sets of conventions do we recover the curved space-time of General Relativity.

We finish this argument with a technical example. The hyperbolic curved space representations of the Universe are normally expressed by the geometrical line element of Riemannian geometry as

$$ds^2 = dt^2 - R^2(t)\left(\frac{dr^2}{1+r^2} + r^2 d\theta^2 + r^2 \sin^2 \theta d\phi^2\right)$$

where

$$R(t) = A(\cosh \psi - 1)$$
$$t = A(\sinh \psi - \psi)$$

By a simple change of coordinates this can readily be transformed into[7]

$$ds^2 = \left(1 - \frac{T_0}{(\tau^2 - \rho^2)^{\frac{1}{2}}}\right)^4 (d\tau^2 - d\rho^2 - \rho^2 d\theta^2 - \rho^2 \sin^2 \theta d\phi^2),$$

that is some multiple of a flat space. The motion of a photon given by $ds = 0$ and therefore follows a straight line in the Euclidean space. A body of mass m_0 follows a geodesic in the Riemannian space

$$\delta \int m_0 ds = 0.$$

In the Euclidean space it no longer follows a straight line but a path given by

$$\delta \int m_0 ds = \delta \int m_0 \left(1 - \frac{T_0}{(\tau^2 - \rho^2)^{\frac{1}{2}}}\right)^2 d\eta = \delta \int m_0 \phi d\eta.$$

The particle's path is determined by the "field" ϕ in the Euclidean space, ϕ is determined by field equations which yield

$$\phi = \left(1 - \frac{T_0}{(\tau^2 - \rho^2)^{\frac{1}{2}}}\right)^2.$$

There is no empirical way of choosing between these alternatives — they predict identical results, yet in the General relativistic representation we say the Universe is curved, in the second that it is flat.

Is space curved? The answer is yes or no depending on the whim of the answerer. It is therefore a question without empirical content, and has no place in physical inquiry.

REFERENCES

1. J. A. Wheeler, *Geometrodynamics*, p. 225, Academic Press, 1962.
2. See the *Leibniz–Clarke Correspondence*, ed. H. G. Alexander, Manchester University Press, 1965.
3. Ernst Mach, *The Science of Mechanics*, 1883, English Translation, Open Court Publishing Co., 1960.
4. See H. Grunbaum, *Philosophy of Space and Time*, 2nd edition, North Holland, 1974.
5. H. Poincaré, *Science and Hypothesis*, English Translation, Dover, 1929.
6. E. A. Milne, *Kinematic Relativity*, Oxford, 1948.
7. I. W. Roxburgh and R. K. Tavakol, *Monthly Notices of the Royal Astronomical Society*, **170**, 599 (1975).

Bruno Bertotti

Professor of Quantum Mechanics at the University of Pavia, Italy.

Worked under the supervision of E. Schrödinger at Dublin Institute for Advanced Studies. Was for a long time active in the contribution of science and scientists to disarmament, and attended Pugwash Conferences on Science and World Affairs.

Scientific interests cover the theory of gravitation, plasma physics and space physics. During recent years particular interests have been in gravitational wave detection and the use of space missions to measure the gravitational field of the Sun.

THE RIDDLES OF GRAVITATION

1. THE PECULIARITY OF GRAVITATION IN THE HISTORY OF SCIENCE

The catalogue of our ignorance has *two*, not one, gates: there is the obvious exit gate, through which questions answered and settled by experimental and theoretical developments march out and disappear into the textbooks and the applications; but there is also a more important, albeit less perspicuous, *entrance gate*, through which *new* riddles come to life in the scientific world. In fact it is an important feature of the history of science (stressed, for example, in the book by T. S. Kuhn, *The Structure of Scientific Revolutions*) that very often the problems set to a scientific community at any given time had been considered beforehand solved or irrelevant; and this process of growing ignorance is usually met with great resistance, since it involves abandoning well-established certainties and efficient formal methods.

After a new area of ignorance has been accepted and defined, its gradual filling up is a fairly standard process, in which the usual interplay between theory and experiments, between hypothesis and verification engages a thriving scientific community; but the appearance of a new riddle is a most delicate and unforeseeable step, like the birth of a new baby in a family: it throws confusion and bewilderment in its plans and opens the way to entirely new developments. This phase deserves careful attention: it is important to know why some places more than others in the established pattern of scientific knowledge are likely to break and to open a gap into a new riddle; and why sometimes it takes such a long time before the riddle becomes really active.

I wish to submit the study of gravitation has served in the past, and is likely to do so in the future, as a *midwife* for the birth of new riddles in the physical sciences. We can quote at least three examples in the past of a feature of gravitation which has triggered a development of much wider scope and consequences to other disciplines. Whether this more advanced level of our field will bear similar fruits in the future, we do not know for certain; but it is important to spot those unanswered problems of gravitation which betray possible, new perspectives. We shall see that, in spite of the precise formal structure and the present excellent agreement with experiments, the theory of gravitation places such deep questions to justify the qualification of "riddle".

2. THE PAST RIDDLES OF GRAVITATION

The first attempt to a dynamical theory goes back to Aristotelis, who distinguishes two kinds of motion, those that have no intrinsic justification ("unnatural"), like the shooting of an arrow, and those which can be understood from the nature of the bodies ("natural motion"). Each body strives to go to its natural place and it will indeed go there, if unimpeded. The basic example of natural motion is fall under gravity: heavy bodies, like stones, belong to the earth and tend to reach it; air and flames belong to the sky and tend to go up. The concept of natural motion is extended to celestial bodies, although they have nothing to do, in this conception, with gravitation. The restricted and basic experience of free fall has suggested a concept which is assumed to be a universal property of things and is applied to the whole philosophy of nature: every being has a "natural" state and will tend to evolve towards it. This interpretation of free fall is the origin of deep riddles: what is "up" and "down", after all?; the strife to deal with them was exceedingly fertile. For example, Aristotelis in his *Physica* noticed that in vacuum there can be no privileged direction of motion ("up" and "down" can be said only in relation to a reference body); hence he concluded that space cannot be empty, since it would not allow natural motion.

The idea of a common law governing both the fall of bodies and celestial motion is due to Newton (1687) and is based upon a *gravitational force of attraction* between two point-like masses inversely proportional to the square of their distance. It had a tremendous success and even nowadays is the basis for the exceedingly accurate predictions of celestial mechanics and for the planning of space navigation of artificial satellites; but it opened up the embarrassing riddle of the "action at a distance" (with the more precise latin words, *actio in distans*). Newton himself was greatly puzzled by this and said: "That one body may act upon another at a distance through a vacuum, without the mediation of anything else . . . is to me so great an absurdity that I believe that no man, who has in philosophical matters a competent faculty for thinking, can ever fall into it." In spite of this riddle, in the eighteenth century the bad example set by gravitation teemed and other important instances of *actio in distans* were established in electricity and magnetism. People tried for a long time to save everything with complicated and unobserved ethereal media, carrying the action from point to point at a finite speed; but at the end modern field theory emerged, according to which a body is supposed to generate in the space around it an abstract quantity, defined mathematically at every event; this quantity — a scalar, a vector, a tensor, etc. — is not directly observable, but determines the force acting upon a test particle. The riddle has served its function: the area of ignorance it had opened up to investigation has been essentially filled up by electromagnetism and allied theories.

3. THE RIDDLE OF GEOMETRICAL SIMPLICITY

The next, fundamental advance was due to Einstein (*Theory of General Relativity*, 1915) and consisted in reducing the theory of gravitation to geometry: the world is a Riemannian, four-dimensional manifold, where the separation between two neighbouring events can be measured by precise clocks.[†] In this manifold privileged lines are defined, as those which connect two events with the shortest (or longest) path; these lines are the possible trajectories of test particles. The gravitational force is hidden in the geometrical structure of the manifold and is described quantitatively by its curvature. Like Newton's, Einstein's theory had a striking and unexpected success: minute deviations from Newton's law of motion have been predicted and verified with a good accuracy (a few per cent) in a large number of experiments. Its simplicity and conceptual beauty have set a standard which has never been met in other physical theories, as one can see from the following points. All physical quantities have a geometrical meaning and can be measured independently with clocks. The dynamical laws for the motion of bodies need not be postulated separately from the field equations which determine the geometrical structure of the manifold, but follow from the latter; this is a much simpler system of postulates than, for example, in electromagnetic theory, where one can, in principle, change the law of motion without changing the field. The large amount of freedom in constructing the possible equations of ordinary field theory contrasts with the almost uniqueness of the field equations for gravitation, severely constrained by geometrical requirements. Finally, the topological properties of the Universe at large are in principle

† See Note A, p. 97.

included in the fundamental physical laws: for example, whether one comes back to the same place or not if he moves straight ahead without ever deviating, follows from the theory in a given solution.

The success of Einstein's theory has produced the "riddle of simplicity", in particular, the riddle of geometrical simplicity: one wonders whether the paucity of postulates and the use of self-contained mathematical structures is indeed a universal key for interpreting and modelling physical experience. An extensive trend of physical investigation was started and shaped by this hope and consisted in studying new, abstract mathematical structures and attempting to endow them with physical interpretation, relying upon experience only as far as fundamental properties are concerned. To quote C. Misner and J. Wheeler (*Annals of Physics*, 2, 525 (1957)): "Two views of the nature of physics stand in sharp contrast: (1) The space time continuum serves only as *arena* for the struggles of fields and particles. (2) There is nothing in the world except empty curved space. Matter, charge, electromagnetism, and other fields are only manifestations of the bending of space. *Physics is·geometry*." These attempts have been particularly intense in the 1940s and the 1950s; one of them, led by Einstein himself and by E. Schrödinger, aimed at formulating a new geometrical theory, to include both electromagnetism and gravitation.[†] Somehow, people felt, classical electromagnetism was inferior to the theory of gravitation, it attained a more superficial level of "reality". Most of these attempts have miserably failed and many physicists were led to rely more upon experimental clues than on aesthetical values in constructing new theories. The riddle of geometrical simplicity — the hope to understand on mathematical grounds why the world is as it is — is still far from fulfilment and keeps its function of stimulating the ingenuity of theoretical physicists.

4. THE RIDDLE OF PROPER TIME

Einstein's theory of gravitation is based upon a mathematical concept — proper time — which is used to define the invariant separation between two events. Every other physical concept in the theory can be derived from it: spatial intervals, the geodesic character of world lines of freely moving bodies, the gravitational field as described by the curvature of the Riemannian manifold, and so forth. In order to measure it one uses a "good clock", usually taken as the best available on the market as far as accuracy and stability; but this hides a deep inconsistency, already pointed out by E. Mach (*The Science of Mechanics*, Dover editions, p.273): "It is utterly beyond our power to measure the changes of things by time. Quite the contrary, time is an abstraction, at which we arrive by means of the change of things. . . . A motion may, with respect to another motion, be uniform; but the question whether a motion is in itself uniform, is senseless. With just as little justice, also, may we speak of an 'absolute time', of a time independent of change. This absolute time can be measured by comparison with no motion; it has therefore neither a practical, nor a scientific value and no one is justified in saying that he knows aught about it. It is an idle metaphysical conception."

This vicious circle is quite clear in Einstein's theory: a clock is needed to test free fall, hence also the dynamical laws on which its functioning itself is based. For the most advanced applications people nowadays use atomic clocks based upon the "hyperfine structure" of spectral lines: their "ticking rate" is given by the frequency of interaction between an electron spin and a nuclear spin in a cesium or a hydrogen atom. This frequency is expressed in terms of fundamental constants by

[†] See Note B, p. 98.

$$\frac{m_p e^4}{h^3} \ \frac{m_e}{m_p} \left(\frac{e^2}{h\,c}\right)^2 ,$$

where h is Planck's constant, e the electron charge, c the velocity of light, m_e and m_p the masses of the electron and the proton, respectively. The good functioning of an atomic clock is based upon the validity of electromagnetism and quantum theory in a small domain; in particular, when the clock is used to measure gravitational fields, one must assume that the dynamical laws in a small region of space are not affected by gravitation. This assumption, called the *Strong Equivalence Principle* (SEP) has observable effects and has been tested with a good accuracy, thus giving a sound, practical basis for the definition of proper time; but violations could occur and, indeed, an important one may follow from the analysis of the lunar motion by T. Van Flandern (see *Scientific American*, February 1976). Leaving aside technical details, he claims to have found that the gravitational "constant" G diminishes with respect to atomic time, with a time scale comparable to the age of the Universe. If this result is confirmed, either we have to alter Einstein's theory (for which G is rigorously constant), or we must assume that one or more fundamental constants entering the above formula undergo a cosmological change, possibly connected with the expansion of the Universe. In this alternative the laws of microscopic physics must change with time (and probably also with space, since the cosmos is not uniform) and do not lead any more to a good proper clock.

Van Flandern's claim can be phrased also by saying simply that atomic time does not agree with "gravitational time", that is to say, the time measured by the revolution of two bodies bound by gravitational force, like the Moon and the Earth; the connection of this with Mach's argument is clear and leads one to think that the very concept of proper time — an absolute entity of metaphysical flavour — is essentially unfit to attack these very important aspects of our experience. One should look for a theory in which *all clocks are treated on the same footing* and physical meaning is given only to their relative rates: there are different and equivalent "times", each related to different interaction mechanisms. In normal conditions, of course, all (well functioning) clocks agree with a very good accuracy and would lead to proper time. The difficulty is, of course, that we know how to describe a dynamical interaction only with respect to a time variable given to us *a priori*. It is clear that such a theory must be very different from general relativity; it might be based, perhaps, on the concept of *conformal invariance*, according to which all the Riemannian manifolds, different only by an arbitrary change of the separation between two neighbouring events, are considered physically identical.

We should mention at this point the deep relationship between the last two "riddles": the reduction of gravitational physics to geometry and the strong equivalence principle. In a geometrical theory the motion of those bodies which are so small and light that do not affect the structure of space-time, is given once for all and cannot depend on their nature and their mass: it follows from the geometry. This assumption is called the Weak Equivalence Principle (WEP). We can show that a violation of SEP may imply a violation of the WEP. Consider, for example, a composite test body made up of two parts, kept together and very near by a force which depends upon the gravitational potential, in contradiction with the SEP. Its binding energy W — the energy required to separate its parts — is therefore also dependent on the gravitational energy and changes from place to place. The theory of special relativity tells us that the binding energy W contributes to the mass m of the body with a term W/c^2; hence the work needed to raise the body in a gravitational field has an additional contribution due to the change in the mass m, which is larger for bodies more strongly bound. The motion of the body then depends upon the ratio W/m and is not given by geometrical properties alone.

The WEP has been verified experimentally with an exceedingly good accuracy; this implies then a (less accurate) test of the SEP for those forces which contribute to the binding energy.

The proper time riddle and riddle of geometrical simplicity call for more and better experiments to provide the theoreticians with clues and constraints. An important, recent result related to it was obtained with the help of the special mirror placed by the American astronauts on the Moon: by sending out to it a very short pulse of laser light and measuring the two-way transit time, it is possible to get the distance between the ground station and the Moon with the surprising accuracy of less than 10 cm. This makes it possible to test various theories for the lunar motion and, in particular, to check if a large, celestial body fulfils the WEP, as Einstein's theory assumes. This test has not, so far, given any discrepance.

5. THE RIDDLE OF INERTIA

In Mach's book *The Science of Mechanics* another criticism of Newton's theory dynamics is raised, which has deep consequences for the physics of gravitation. When in the second law one equates the product of (inertial) mass by acceleration to the force, he must make specific assumptions concerning the frame of reference: the law is usually founded upon the empirical basis of the distant galaxies being, on the average, at rest. This works very well in practice: it determines the zero of the angular rotation of our coordinates with an accuracy of about 1″ per century. But it raises the theoretical problems, how is the frame of reference determined if distant galaxies would rotate differently in different directions, and why do they affect in such an essential way the local dynamics?

This connection was indeed acutely present in Einstein's mind when he was constructing the theory of general relativity. There a single entity — the metric field — determines the dynamics and the measurements of space and time, in particular the frame of reference; the metric on turn depends upon the distribution of matter in the Universe. In Einstein's theory it can be expected that, for example, the rotation of a shell of matter produces in this interior a dynamical change similar to the change induced by a rotation of the frame of reference. This typical Machian effect is indeed a consequence of the theory of general relativity (Lens Thirring effect). A large amount of work has been done along this line; we should also quote the papers by D. Sciama, D. Lynden Bell and others who have reformulated Einstein's field equations in a way in which the matter distribution determines *uniquely* the metric. In the conventional theory of general relativity it is necessary to supplement the matter distribution with the prescription of the conditions which the metric must fulfil at infinity. They are equivalent to conditions arbitrarily imposed in the frame of reference and constitute a distinctly non-Machian feature of general relativity.

It has been pointed out by MacCrea (*Nature* **230**, 95 (1971)) that the adoption of special (and *a fortiori*, of general) relativity reduces drastically Mach's problem. In special relativity no material body can have a velocity larger than c; hence if the Universe has a radius R its largest rotation velocity is c/R, an exceedingly small quantity. Similarly, the largest acceleration we can admit is of order c/T where T is the age of the Universe. Therefore the assumption that distant matter obeys the rules of special relativity implies assigning with a very good accuracy the inertial frame of reference. (In general relativity the conclusion is not so strict, but rather elaborate and *ad hoc* constructions are needed to shirk it: for example, a gravitational field could twist the rays of light so as to make matter at rest far away appear rotating to a local observer.) *Accepting the theory of relativity, therefore, means excluding* a priori *a rational explanation of the most surprising and very accurate agreement between the inertial frames of reference* (in which there is no centrifugal force) *and the standard of rest of distant galaxies.*

Perhaps this was a reason why Mach felt reluctant to accept Einstein's theory of relativity; in this extreme view it should be considered only as a (good) approximation which one arrives at when the dynamical behaviour of the Universe is taken into account and the frame of reference is thereby determined. The usual sequence of theories

1. Newton's
2. Special Relativity
3. General Relativity
4. Machian General Relativity

should perhaps be replaced by

1. Newton's
2. Machian theory of gravitation
3. Special Relativity as an approximation
4. General Relativity.

Considering the immense success of and the lack of any serious challenge to Special Relativity in more than 70 years, it takes a good deal of courage from anybody who tries to follow this unorthodox course. A rudimentary, but interesting attempt along this direction was made recently by J. Barbour and B. Bertotti (*Nuovo Cimento*, 38B, 1 (1977)). They have formulated a dynamical theory of N point-like masses in which only observables appear: an arbitrary time variable and the $\frac{1}{2}N(N+1)$ distances between them; when the masses are divided in two groups, one far away and all around and a local one, one recovers Newton's equations of motion for gravitating bodies. Thus, Galilei invariance follows from an appropriate cosmology.

In an Encyclopaedia of Ignorance it is certainly appropriate to take the risk to mention such an unusual and unconventional point of view; only (long) time will show the real depth of this last riddle of gravitation, if there is any.

Notes

A. It is both possible and interesting, even for a layman, to know the technical details of this mathematical structure. A four-dimensional manifold is a continuum of "points" labelled by four real numbers x, y, z and t. Each point corresponds to an event of space-time. The qualification "Riemannian" (from the mathematician B. Riemann) is given to a manifold in which to every pair of neighbouring events (x, y, z, t) and $(x + dx, y + dy, z + dt, t + dt)$ is associated a real number, quadratic function of the small increments dx, dy, dt and dt of the coordinates:

$$\Phi = g_{xx}\, dx^2 + 2g_{xy}\, dx\, dy + 2g_{xz}\, dx\, dz + 2g_{xt}\, dx\, dt$$
$$+ g_{yy}\, dy^2 + 2g_{yz}\, dy\, dz + 2g_{yt}\, dy\, dt$$
$$+ g_{zz}\, dz^2 + 2g_{zt}\, dz\, dt$$
$$+ g_{tt}\, dt^2 .$$

This is a generalization of the distance

$$\sqrt{(dx^2 + dy^2 + dz^2)}$$

between two points in ordinary space with Cartesian coordinates (x, y, z) and $(x + dx, y + dy, z + dz)$ (Pythagoras' theorem). When Φ is positive, its square root gives the time interval between the two events; when Φ is negative, $\sqrt{-\Phi}$ is the spatial distance between two events. The length of line segment is obtained by summing the contributions $\sqrt{\Phi}$ (or $\sqrt{-\Phi}$) of each its infinitesimal segment.

Every information about physics is contained in the ten function of x, y, z and t, g_{xx}, g_{xy}, g_{xz}, g_{xt}, g_{yy}, g_{yz}, g_{yt}, g_{zz}, g_{zt} and g_{tt}, collectively denoted as the "metric"; in particular, the "curvature" of the manifold indicates the amount of matter present at each event and describes its gravitational field.

B. In this theory the array of the metric field is completed as follows:

$$
\begin{array}{cccc}
g_{xx} & g_{xy} & g_{xz} & g_{xt} \\
g_{yx} & g_{yy} & g_{yz} & g_{yt} \\
g_{zx} & g_{zy} & g_{zz} & g_{zt} \\
g_{tx} & g_{ty} & g_{tz} & g_{tt}.
\end{array}
$$

The six added quantities correspond to the electric field (three components) and magnetic field (three components); the sixteen functions of space-time thus obtained form a single entity and describe in a unified manner gravitation and electromagnetism.

Gary Benzion, Cornell Daily Sun

Thomas Gold, F.R.S.

Director, Center for Radio-Physics and Space Research and John L. Wetherill Professor, Cornell University.

Co-author with Hermann Bondi of the *Steady-State Theory of the Origin of the Universe*.

Worked in Cambridge at the Cavendish Laboratory and at the Royal Greenwich Observatory. Formerly Professor of Applied Astronomy at Harvard University.

Published research has been in the areas of astronomy, physics and biophysics.

RELATIVITY AND TIME

We think we have discovered, by the most direct observation, that there exists a three-dimensional space and a uniform flow of time. However, I wonder whether this is a good or indeed a true representation of reality.

Relativity Theory has given us the details of the relation between space and time. It has also taught us that all future events to which we may ever have access can be seen, from a suitably moving observer's viewpoint, as simultaneous with the present (or arbitrarily closely so). The flow of time is abolished, but the relationship between events is still unchanged.

Perhaps three-dimensional space is only one of a variety of ways in which our environment can be described. What we really know in our minds is, after all, only an algebraic system, and we note that it has a close correspondence to a particular type of geometry.

The *flow* of time is clearly an inappropriate concept for the description of the physical world that has no past, present and future. It just is. Perhaps space is also an inappropriate concept, and an algebraic system can be invented, whose rules are simpler and yet describe the relations between events. Perhaps this will allow us to see Relativity Theory as an obvious structure, not as something complex and difficult. Perhaps the laws of physics will have a place in this, not be added as an afterthought to a subjective and possibly capricious definition of the universe in terms of the strange notions of space and time.

A. J. Leggett

Reader in Theoretical Physics at the University of Sussex.

Educated at Oxford and carried out research at the Universities of Illinois and Kyoto.

Main current field of research is the macroscopic manifestations of quantum-mechanical behaviour shown by superconductors, superfluids and possibly other systems. Recently has been particularly involved in research on the new ultra-low-temperature phases of liquid helium. He has also maintained a lasting interest in the fields of intersection of physics and philosophy, and regularly teaches courses in this area.

THE "ARROW OF TIME" AND QUANTUM MECHANICS

In those exciting but frustrating fields of knowledge, or perhaps one should say ignorance, where physics tangles with philosophy, the difficulties usually lie less in finding answers to well-posed questions than in formulating the fruitful questions in the first place. The attempts which follow to do this for one particular area could charitably be described as at best quarter-baked, and may well reflect the ignorance and confusion of the author rather than that of the scientific community as a whole.

Most natural scientists probably have a deeply ingrained belief that it should be possible to give a complete description of the laws of nature without explicit reference to human consciousness or human intervention. Yet at the heart of physics – for long the paradigm for natural science – lie two problems where this assumption is still subject to furious debate. One is the question of the "arrow of time" or more correctly the apparent asymmetry of physical processes with respect to time,[1, 2] the other the problem of measurement in quantum mechanics.[3] In this essay I speculate on the relationship between these at first sight disconnected problems and the vast areas of human ignorance which may possibly lurk behind them.

At the level of classical mechanics, electrodynamics and (with one minor proviso which is unimportant for present purposes) elementary particle physics, the laws of physics as currently accepted indicate no "preferred" direction of physical processes with respect to time: in technical language, they are invariant under the operation of time reversal. Crudely speaking, every process in (say) mechanics which is possible in one direction is also possible in the reverse direction: for example, if a film were taken of a system obeying the laws of Newtonian mechanics without dissipation (such as is constituted, to a very good approximation, by the planets circling the Sun), and if such a film were then projected backwards, it would be impossible to tell this from the film alone. Moreover – a related but not identical point – the laws of physics at this level give no support to the idea that the present (or past) "determines" the future rather than vice versa. Indeed, while it is certainly true that from a knowledge of the positions and velocities of the planets at some initial time we can in principle calculate the values of the same quantities at a later time, the converse is equally true: from the values at a later instant we can in principle calculate the earlier values just as accurately. Thus, within the framework of mechanics alone, the idea that the earlier events "cause" later ones rather than vice versa is a (possibly illegitimate) importation of anthropomorphic concepts into the subject: and the same is true for electrodynamics and elementary particle theory.†

Any extrapolation of these ideas to the whole of physics would, however, obviously run violently counter to common sense. It is a matter of observation (not of interpretation!) that there exist very many spontaneously occurring processes in nature whose time inverses do *not* occur spontaneously, for instance the melting of ice in a glass of warm water or the gradual loss of energy of a bouncing ball. In such cases a film of the process run backwards would be immediately recognizable as such. Moreover, most people, at least among those brought up in the modern Western intellectual tradition, would certainly say that the present (and past) can influence the future but not vice versa. Perhaps, however, it is worth noting even at this stage that this has not always been the dominant belief and may indeed even now not be the dominant one on a world scale: and while, historically speaking, these societies whose beliefs could be crudely called teleological or fatalistic have not usually embraced natural science in

† It is a misconception (though one surprisingly widespread among physicists) that experiments in elementary particle physics have "proved causality" in a sense which would determine a unique direction of the causal relationship in time. On close inspection it turns out (as usual in such cases) that in the interpretation of the experiments the "direction of causality" is already implicitly assumed.

the sense in which we understand it, it is worth asking whether this correlation is a necessary one.

Let us return for the moment to the occurrence in nature of "irreversible" processes, that is, those whose time inverses do not occur spontaneously, and which could therefore apparently be used to define a unique "direction" of time. Such processes are, of course, part of the subject matter of thermodynamics and statistical mechanics, and the conventional explanation of the apparent asymmetry despite the time-symmetry of the underlying microscopic laws goes, very crudely, as follows: if a system is left to itself, its degree of disorder (technically, its entropy) tends to increase as a function of time. To use an often-quoted analogy, if we shuffle a pack of cards we will almost always make it more disordered: if we start with the cards arranged in "perfect order" (ace, king, queen of spades on the top, etc.) we will almost inevitably end up, after shuffling, at a less ordered distribution, but it is extremely unlikely that starting from a "random" pack we would end up, by the ordinary process of shuffling, at a perfectly ordered distribution, and indeed any player who achieved such a result would almost automatically be suspected of cheating. Such an increase of disorder (entropy) seems at first sight to be naturally asymmetric in time and hence to define a unique "direction".†

Let us assume for the moment that the ascription of more "order" to the unshuffled ("perfect") distribution than the "random" one is an intuitively transparent operation which involves no implicit anthropomorphic elements and moreover that it can be made satisfactorily quantitative (cf. previous footnote). Actually, the first of these assumptions at least is by no means unproblematical (what is a pack of cards *for*? Would Martians recognize the perfectly ordered pack as such? etc.) but it is not what I want to discuss here. Rather I want to focus on a difficulty which is well known and has been discussed very thoroughly in the literature,[1, 2] namely that in any purely physical process, the underlying dynamics of the system is time-reversible as mentioned earlier, and therefore that disorder will tend to increase in *both* directions: in fact, if we *knew* for certain that a given pack of cards had been shuffled "at random" and nevertheless found that at the time of observation it was completely ordered, we could legitimately conclude not only that a few minutes later, as a result of further shuffling, it would be more disordered, but also that a few minutes *before* the observation it had also been less ordered, and that we just happened to have caught it at the peak of a very unlikely statistical fluctuation. Now in practice, of course, if we found a perfectly ordered pack *of whose history we knew nothing*, we would conclude nothing of the sort, but rather that it had been deliberately prepared in order by purposeful human intervention. Thus, the apparent asymmetry of the increase of disorder with time in a case of this kind is a consequence of the fact that human beings can "prepare" highly ordered states, whereas they cannot "retropare" them: that is, they can set the initial conditions for a system at time $t < t_1$, and then let the laws of physics take their course undisturbed between, say, t_1 and t_2. If the initial conditions correspond to a highly ordered state, the degree of disorder is practically bound to increase in time despite the time-symmetry of the underlying laws. The inverse procedure is impossible, at least according to common sense: to determine the state of the system at the final instant, t_2, we would have to intervene *before* t_2, that is precisely during the interval when the laws of physics were supposed to be taking their course without outside intervention. So, in the end,

---

† The analogy as presented is, of course, very much oversimplified. In fact, if the "disorder" or "entropy" of a pack of cards is to correspond to the concept used in statistical mechanics, it should be a characteristic not of a particular arrangement of the cards ("microstate") but of a class of arrangements ("macrostate"), having some gross property or properties in common, e.g. a particular value of the number of pairs of neighbouring cards which are of the same colour. For details see, for example, ref. 4. However, for the limited purpose for which I need the analogy here it is unnecessary to go into these complications.

the apparent asymmetry implied by the second law of thermodynamics (increase of disorder) turns out to be intimately related to the fact that we can only affect the future — that is, to the second paradox indicated above, namely the fact that while we have a strong sense of the "direction" of causality in the macroscopic world, microscopic physics supplies no obvious basis for such an idea. Perhaps it is now becoming apparent in what way human consciousness, and human intervention, may at first sight at least be involved in the problem of time asymmetry even within the realm of physics.

At this stage it is as well to digress a moment to dispose of one matter which is not directly relevant to the present argument. It goes without saying that there are plenty of irreversible processes in nature where no human intervention is or could possibly have been involved. In such cases the asymmetry in time is generally and plausibly ascribed to the obvious asymmetry in the natural environment which arises from the fact that the Sun is radiating energy outwards rather than sucking it in "from infinity" — technically, it is a source rather than a sink of radiation. This fact is in turn usually related to the so-called "cosmological" time asymmetry — the fact that (according to most current theories) the Universe as a whole is expanding rather than contracting. This is a topic which itself involves some fairly deep conceptual problems,[2] but it does not directly affect the argument as presented above. However, we shall encounter it again later.

Turning back to the question of situations involving human agency, we have just seen that the apparent asymmetry of such situations in time is a consequence of the inability of human beings to affect the past. The question I now want to raise is: Does physics itself, directly or indirectly, via, say, biological considerations ultimately based on physics, provide any obvious reason for this inability? Or, indeed, for our somewhat related inability to "remember the future"? Such questions might seem at first sight absurd, and indeed may be made so by any of a number of quite natural misinterpretations: if, for instance, we were to interpret the second question as asking why it is the past we remember *rather than* the future, we risk inviting a reply in purely linguistic terms, namely that the past is, by linguistic convention, the "direction" which we can remember. To focus the discussion, therefore, it may be helpful to ask two more specific questions: Does physics by itself forbid: (a) the hypothesis that intelligent beings in a distant part of our Universe (Daleks for short) should have a sense of the "direction" of time which is reversed with respect to ours, (b) a limited degree of genuine precognition among us ordinary human beings?

Hypothesis (a) is only interesting (and indeed perhaps only meaningful) if, apart from their possibly inverted sense of time, the nature of Dalek "life" is sufficiently close to our own to be recognizable as such.† In that case we should have to ask, what are the conditions on our own environment which (for instance) allow an organism to differentiate in one direction in time rather than the other: if, as is commonly accepted, a necessary condition is the constant input of radiation energy from the Sun, then our "biological" and hence our (overall) "psychological" arrow of time is a consequence of the "cosmological" arrow. Since the Daleks are inhabitants of the same universe, they share this latter arrow and their sun is also presumably a source rather than a sink of radiation. It would then follow that hypothesis (a) is excluded — although it is by no means trivial to fill in the details of the above argument.[6]

However, even if we accept this conclusion it is by no means obvious (to me at least) that we should then exclude hypothesis (b) without further discussion. Indeed, to an unprejudiced observer, the evidence,[7] anecdotal as it inevitably is, for a strictly limited degree of

† The question of whether, and how, we could recognize "time-inverted" beings as intelligent[5] is a fascinating one, though unlikely to be of practical interest to space biologists!

precognition may be thought quite impressive. (Ability to "affect the past" would presumably be considerably more difficult to recognize!) And what I want to suggest is, that at least in the absence of a much more detailed understanding of the workings of the human brain than we at present possess, it is *not* entirely obvious that the laws of physics, even when combined with the given *overall* direction of biological process, exclude any possibility of genuine precognition over fairly small distances in time — or, by the same token, of a very limited ability to "affect the past". Needless to say, such a possibility, were it to exist, would have profound implications not only for philosophy but also for our view of physics itself.

Let us now turn to the second, at first sight unrelated, area of physics where human consciousness is sometimes believed to play a special role, namely the theory of the measurement process in quantum mechanics.[3] In the standard formulation of quantum mechanics,[8] one talks strictly speaking not about a single system, but about an "ensemble" (class) of identically prepared systems, and describes such an ensemble by a "wave function", knowledge of which enables us to predict, *purely statistically*, the probability of various outcomes of a given measurement. Consider, for instance, the measurement of some quantity A which can take one of a finite number of discrete values† $a_1, a_2, \ldots, a_i, \ldots$. The wave function then allows us to predict unambiguously the *probability* p_i of getting a specified outcome a_i of the measurement on a system taken at random from the ensemble. Yet the wave-function description certainly cannot be replaced by a description in which a fraction p_i of the systems forming the ensemble are said to be *in* the state corresponding to a value a_i of the quantity A: the two descriptions would in fact give in general quite different predictions of the outcome of an experiment in which a quantity different from A is measured, and to date at least the predictions of the wave function description seems to be in good argument with experiment while the predictions of the rival description are not.

One is therefore apparently forced to say that until the quantity A is measured it *does not have a definite value* for any individual system in the ensemble. (It should be emphasized, again, that it is not just a question of us not knowing the value: if this were the case, the description rejected above would be adequate.) On the other hand, according to the standard prescriptions of quantum mechanics, the moment that the quantity A is measured the description of the system undergoes a discontinuous change: in fact the wave function changes abruptly so as to accommodate the information that A now "has" the value we have just measured (the so-called "reduction of the wave packet"). One is then faced with a dilemma: *either* (a) the wave function characterizes some physical property (which we at present find difficult to interpret intuitively) of each individual system: *or* (b) it is merely a shorthand for the statistical properties of the ensemble (and thus, naturally, has to be rewritten as soon as we obtain additional information about any particular system).‡ The difficulty with interpretation (a) is that in all but the very simplest cases it is hard to see how the assumed physical property can suddenly and discontinuously change when a measurement is made;[9] for instance, if a single electron (or photon) is diffracted through a narrow slit, its wave function spreads out over a wide area — and so, presumably, there is actually some kind of (presently unknown) physical disturbance over the whole of this area. Yet the moment that the position of the particle is "measured" (e.g. by observing the highly localized flash it makes on a scintillating screen) the

---

† Such a quantity might be, for example, the projection of the particle's intrinsic angular momentum on a given axis. Of course, the very fact that only discrete values of such a quantity are allowed is a consequence of the quantum theory and would not occur in classical mechanics.

‡ Technically speaking, by obtaining more information about a given system we assign it to a new ensemble.

wave function is supposed to contract into a very small volume near the position of the flash; and so, presumably, does the associated physical disturbance. Indeed, if we take the prescription literally, the contraction should actually take place at a rate exceeding the speed of light, thereby violating the canons of special relativity. With interpretation (b) above, in which the wave function represents nothing physical but is simply a statistical device resulting from our ignorance, the difficulty is precisely to see why the statistical description rejected above does not work. Faced with this dilemma, the majority of physicists have embraced, at least implicitly, the so-called Copenhagen interpretation[10] (more correctly, non-interpretation) of quantum mechanics according to which it is meaningless to ask questions, awkward or otherwise, about the meaning of the wave function, since it is simply and solely a mathematical interpolation enabling us to infer from a recipe for the preparation of a particular ensemble to the probability of various outcomes of possible experiments on it.[11]

While most practising physicists find this (non)-interpretation quite satisfactory in the context of their everyday use of quantum mechanics, it has become increasingly recognized over the last 20 years or so that the awkward conceptual problems it raises are not going to be exorcised so easily. Indeed, probably the single issue which divides physicists most deeply at present is the extent to which these problems cast doubt on the claims of quantum mechanics to constitute in some sense the ultimate description of physical reality. Let me sketch very briefly the nature of just one of these problems.[3] The notion of "measurement" clearly plays a special role in quantum mechanics, since the wave function is supposed to change continuously and causally between measurements but to jump (collapse) discontinuously and acausally as soon as a measurement is made.[12] This prescription, however, is ambiguous, since quantum mechanics contains no instructions for deciding exactly when a "measurement" has been made. In fact, it is possible to argue that what we normally call a measurement is nothing more than interaction with a (usually man-made) device such as a Geiger counter which itself consists of atoms and molecules, in which case it should itself be described by a quantum mechanical wave function; but if so, the device itself will in general not possess a definite value of all its physical quantities (including the readings of dials, etc.) until these are themselves "measured", and so on in an infinite regress. Indeed, some authors [13] have argued that the only satisfactory way to terminate this regress is to allow the "measurement" to be made only when the reading of the dials (say) is registered by a human mind, thus introducing human consciousness into the theory as an extra-physical ingredient. Clearly the paradox here is somewhat reminiscent of the one we encountered earlier in the context of time-reversibility: to enable thermodynamics to provide a unique "direction of time" we were forced in the end to assume that human beings can affect the future but not the past, an assumption which (apparently) cannot be justified within physics itself: to formulate quantum mechanics at all, we have to introduce the idea of measurement, a notion which (apparently) cannot be defined without internal inconsistency in physical terms alone.

In the face of this and other difficulties, some physicists have speculated that quantum mechanics may actually only be an approximate description of reality, the true description involving a consideration of a sub-quantum level characterized by some variables which at present we are unable to detect experimentally (the so-called "hidden variables"). Such a description, it is hoped, could be completely causal in character, and would if suitably constructed reproduce the quantum-mechanical results for existing types of experiment while avoiding the conceptual difficulties concerned with the problem of measurement; moreover, it might in principle predict results different from quantum mechanics under conditions more stringent than those hitherto attained.[14] A model of such a theory has been explicitly constructed[15] for a simple situation involving spin measurements on a single particle (or ensemble of such), thus finally disproving the widespread but erroneous belief[16] that no hidden-variable theory could reproduce all the results of quantum mechanics even for such a simple system. At first sight, therefore, it looks as if there are no very fundamental *a priori* objections to the idea of such a "hidden-variable" theory.

However, one of the most interesting and surprising developments in fundamental physics in the last decade or so has been the demonstration that the position is quite otherwise as soon as one considers a slightly more complicated situation, namely spin measurement (or the equivalent) on two systems which have interacted in the past but are now very far apart in space (so that, within the framework of presently accepted ideas, they should not influence one another). Let us assume (A) that the hidden-variable theory is such that after the two systems separate, each separately has a state described by its own hidden variables (which we do not, of course, know and which may be strongly correlated, in the usual statistical sense, with those of the other, remote, system); and moreover, (B) that the outcome of any measurement on (say) system 1 is determined only by the state of that system and by the properties of the apparatus set up to perform measurements on it, but not (for instance) by the variables of system 2 or the properties of *its* associated apparatus. (Such a hypothesis seems entirely natural since the two systems, and the two associated pieces of apparatus, are very distant from one another in space.) A theory having these properties is referred to as a "local" hidden-variable theory. There is now a remarkable theorem[17] which states that *no local hidden-variable theory can reproduce all the results of quantum mechanics.* Moreover, with some trivial modifications, the theorem can be strengthened to apply more generally to *any* "local" theory, that is any theory for which assumption (B) above is true and in which (A) is replaced by the more general assumption that there exists a description of the state of each system separately after they have ceased to interact. (In quantum mechanics no such description exists, which itself leads to a well-known paradox.[18] This theoretical conclusion has now been complemented by an experiment[19] which deliberately provided a situation in which the quantum mechanical predictions could not be reproduced by any "local" theory: the quantum predictions were, nevertheless, found experimentally, thus demonstrating fairly conclusively that no "local" description of nature can be correct.

It should be stressed that this conclusion at first sight runs violently counter to common sense, since it says that (at least in certain circumstances) a system cannot even be *described* individually and in isolation, even though it may be spatially separated from all other matter in the Universe. Indeed, if we were to take the argument to its logical conclusion it would seem to say that we can *never* describe any system in isolation, since it must have interacted with something in the past, however long ago!

It is possible to avoid this somewhat unpalatable conclusion if one is prepared to modify one or two of the "common-sense" assumptions embodied, perhaps implicitly, in (A) and (B) above in a sufficiently radical way. For instance, a possible hypothesis[20] is that by setting up the apparatus designed to measure the properties of system 1 one may, in some way not at present understood, affect the physical conditions prevailing in the region where system 2 is and thus the results of measurements made on it. Such a hypothesis, if it is not to violate the special theory of relativity, could in principle be tested by setting up the apparatus "at the last minute", so that there is no time for any signal to be transmitted to system 2.

However, to my mind a more intriguing possibility, and one which at last makes the promised contact with the first part of this essay, is that the "direction of causality" might in some sense be violated in this type of experiment[21] (and, perhaps, more generally in quantum measurement processes). In other words, instead of regarding the initial state of the system (whether described by hidden variables or not) as determining the outcome of measurements made on it, we might regard the outcome of the measurements as determining, at least partly, the initial state. This is perhaps somewhat more plausible in a hidden-variable picture (or some other theory which seeks a "sub-quantum-mechanical" level of reality) since a (temporary) "backward" interpretation of causality at the sub-quantum level might not necessarily conflict with the usual "forward" interpretation of the level of quantum mechanics, nor produce results which are clearly incompatible with the known initial conditions. It should be noticed, by the way, that the typical times involved in experiments of this type are usually extremely small by macroscopic standards (usually of the order of 10^{-9} sec) and indeed are probably less than the shortest microscopic "relaxation time" for the irreversible processes likely to be relevant here. It is therefore perhaps not so unthinkable that phenomena of a type (at first sight) unknown on the time-scale appropriate to the macroscopic world might exist over such short intervals.[22]

Speculating even more wildly, one might hope that if anything remotely resembling this proposal were to be true, it would not only resolve the "measurement paradoxes" of quantum mechanics at the atomic level, but also provide the microscopic basis for those phenomena, if they really exist, which *do* violate the "sense of time" on a *macroscopic* time-scale — such as precognition. That the time scale involved here is so many orders of magnitude longer (perhaps minutes or hours) is perhaps not so strange if one considers that the human brain, regarded as a physical system, certainly possesses a degree of complexity very many orders of magnitude greater than any of the instruments used in physics. On the other hand, it could equally well be that the questions of microscopic and macroscopic violations of the sense of time may turn out to be essentially unrelated, as seems to be the case with the violations of left-right symmetry at the levels of elementary particles and of biology.[23]

The above speculations may seem to be (and no doubt are) vague to the point of irresponsibility. Nevertheless, I do strongly suspect that if in the year 2075 physicists look back on us poor quantum-mechanics-besotted idiots of the twentieth century with pity and head-shaking, an essential ingredient in their new picture of the Universe will be a quite new and to us unforeseeable approach to the concept of time: and that to them our current ideas about the asymmetry of nature with respect to time will appear as naïve as do to us the notions of nineteenth-century physics about simultaneity.†

ACKNOWLEDGEMENT

I am indebted to Dr. Paul Davies for a very helpful discussion and criticisms of the original manuscript.

---

† For a deeper discussion of many of the questions raised here, as well as many recent references, see O. Costa de Beauregard, *Foundations of Physics* (in press).

1. For general discussions of the problem of time asymmetry, see H. Reichenbach, *The Direction of Time*, University of California Press, Berkeley, 1971, and
2. P. C. W. Davies, *The Physics of Time Asymmetry*, Surrey University Press, London, 1974.
3. For a general introduction to the problem of measurement in quantum mechanics, see B. d'Espagnat, *Conceptual Foundations of Quantum Mechanics*, 2nd edition, Benjamin, New York, 1976.
4. R. Kubo, *Statistical Mechanics*, North-Holland, Amsterdam, 1967, chap. 1.
5. Cf. N. Wiener, *Cybernetics*, Wiley & Sons, New York, 1948, p. 34; *The Nature of Time*, edited by T. Gold, Cornell University Press, Ithaca, N.Y., 1967, pp. 140-2.
6. Cf. H. F. Blum, *Time's Arrow and Evolution*, Princeton University Press, Princeton, N.J., 1968.
7. See, for example, J. B. Rhine, *The Reach of the Mind*, Faber & Faber, London, 1956, chap. 5.
8. L. Eisenbud, *Conceptual Foundations of Quantum Mechanics*, van Nostrand Reinhold, New York, 1971.
9. This point is emphasized in (e.g.) L. E. Ballentine, *Rev. Mod. Phys.* 42, 358 (1970).
10. N. Bohr, *Phys. Rev.* 48, 696 (1935).
11. For a more sophisticated version of the "Copenhagen interpretation" see H. Reichenbach, *Philosophic Foundations of Quantum Mechanics*, University of California Press, Berkeley, 1944.
12. For an alternative interpretation which avoids this "collapse" (but involves other difficulties) see H. Everett III, *Rev. Mod. Phys.* 29, 454 (1957).
13. F. London and E. Bauer, *La Théorie de la mésure en mécanique quantique*, Hermann et Cie, Paris, 1939; E. P. Wigner, *Am. J. Phys.* 31, 1 (1963).
14. For a forceful statement of this point of view, see D. Bohm, *Causality and Chance in Modern Physics*, Routledge and Kegan Paul, London, 1957.
15. D. Bohm and J. Bub, *Rev. Mod. Phys.* 38, 453 (1966).
16. This belief goes back to a result of J. von Neumann, *Mathematical Foundations of Quantum Mechanics*, Princeton University Press, Princeton, N.J., 1955.
17. J. S. Bell, *Physics,* 1, 195 (1964-5).
18. A. Einstein, B. Podolsky and N. Rosen, *Phys. Rev.* 47, 777 (1935).
19. S. J. Freedman and J. F. Clauser, *Phys. Rev. Letters* 28, 938 (1972).
20. D. Bohm in D. R. Bates (ed.) *Quantum Theory*, vol. III, Academic Press, New York, 1962, p. 385.
21. To the best of my knowledge this possibility was first pointed out in the present context by O. Costa de Beauregard, *Revue Internationale de Philosophie*, no. 61-62, 1 (1962); *Dialectica* 19, 280 (1965).
22. Cf. the problem of "pre-acceleration" in electrodynamics (Ref. 2, p. 125), where the relevant time is of order 10^{-23} sec.
23. But see T. L. V. Ulbricht, *Q. Rev. Chem. Soc.* 13, 48 (1959).

C. J. S. Clarke

Lecturer in Mathematics at the University of York.

Studied mathematics and general relativity at Christ's College, Cambridge, and was a research fellow at Jesus College, Cambridge, until 1974.

Recent work has been on the structure of singularities in general relativity and on the foundations of quantum theory. This extends into his more fundamental interest in the rediscovery of the intellectual content of contemplative Christianity, an enterprise pursued in conjunction with "the Epiphany Philosophers" at Cambridge.

THE HINTERLAND BETWEEN LARGE AND SMALL

1. PROLOGUE

I shall discuss an area where there is less than ignorance: where knowledge is negative, in that what we think we know must necessarily be false: the region of overlap between microphysics and macrophysics. Likening the growth of knowledge to the progressive construction of an atlas of our world, imagine two cartographers setting out to map the Pacific, one starting from the West and the other from the East, each one ignorant of latitude and longitude but using his own empirical coordinate system centred on his homeland. The initial gulf of ignorance between them would dwindle away until, finally, the maps overlap. In the region of overlap the maps would look different because of the use of two different coordinate systems. But a geometric transformation could convert from one to the other, so that it could indeed be said that the entire Pacific had been charted.

Now imagine their confusion if, in their common region, no such conversion were possible. Suppose that one map held a certain island to be mountainous, the other map flat. Was it really the same island? Did it change according to who observed it? Or had the entire process of map-making been in error all the time? Worse than ignorance, all that had previously been thought knowledge would be thrown into question.

Such, I claim, is the situation with micro- and macrophysics. As the two disciplines extend their maps outward towards each other, each quite satisfactorily, the less likely it becomes that any reconcilable overlap will be achieved unless one or the other undergoes a radical change. And the situation is revealed as crucial in the realization that it is just in the domain of possible overlap, the domain of threatened contradiction, that all the basic processes of life as we know it take place.

The arguments leading to this conclusion are simple ones, based on the general structures of classical and quantum physics. References and some more technical points are postponed to the notes at the end.

2. THE SMALL

(a) In Support of Realism

The theory of the small means, in modern physics, quantum theory — about whose nature and content no two physicists agree. The poles of the conflict correspond to the terms of what is perhaps the most basic classification of the psychology of Western thought, into nominalist and realist.[1] The first type is characterized by the tendency to see all general terms, or sometimes all terms whatever, as mere shorthands for specific events or experiences; the second sees them as pointers to structures that have some kind of "real" existence independent of our observing them. From the nominalist tendency springs operationalism, utilitarianism[2] and the approach to physics which regards theoretical concepts as merely parts of algorithms enabling one to predict successfully the results of experiments. From the realist comes a Platonic view of the world and a use of theories in which mathematical terms are seen as standing for real parts of the furniture of the Universe.

If science is nothing but the prediction of future dial-readings from present ones, there is no terrain to be mapped and so no map. I intend to reject such an extreme form of the nominalist tendency, and shall presume that scientific theories are *about* something, other than the experiments set up to test them. Clearly, all manner of questions are begged in saying this: a large part of philosophy, starting with Kant, has been concerned with just this question. This philosophy warns us, if the history of science were not warning enough, against being seduced by cartographical metaphors into supposing that we can usefully assess a theory by comparing it with some hypothetical "real world". Yet without some measure of realism there is neither knowledge nor ignorance, only greater or less success in prediction. And without a commitment to something underlying the formulae, some belief in a terrain which the map depicts, few steps in science would be possible.

The resolution of the conflict between large and small is to be sought here. Complete nominalism dissolves the conflict, but with it the possibility of knowledge; naïve realism provides satisfying pictures of the world, whose definiteness, however, renders the opposing views irreconcilable.

(b) "States" and "Collapse"

For quantum theory, the key term to place in the nominalist-realist spectrum is "state". The theory then tells us how to express everything in terms of relations between states, starting from the basic assumption that the set of all states has the same structure as the set of all complex one-dimensional subspaces of a complex infinite-dimensional Hilbert space.[3]

A *state* (or, more specifically, a state of an atom or particle) might be specified by what one has done in the past in order to set up some microscopic system; or by what results one can get, and with what probabilities, if one performs certain subsequent operations on the system. If "state" means nothing more than this[4] (the nominalist position) then quantum mechanical theories are nothing more than statements about connections between setting-up-actions and observing-actions. We must go on to ask, is there anything more that "state" can signify?

We customarily think of a state *of something*: that is, we imagine a substratum which possesses certain properties which constitute the state at any given time. In the realm of particle physics this idea is probably untenable, since particles can appear and disappear leaving no fixed substratum.[5] Therefore we must weaken the concept to that of the *state of affairs* maintaining at a given time[6] in some region of space — a weaker concept, but not so weak that we renounce all explanatory power. In the last section I shall explore the possibility of yet further weakening.

This now leads us to the central distinction between the physics of the large and the physics of the small. Macroscopically, so long as a region continues to be designated unambiguously, then a state of affairs can always be ascribed to it, a state which changes more or less continuously and at any rate in accordance with laws which generally produce a fairly predictable continuous change. But the assumptions that a state always has been defined and always will be defined is superfluous to most microphysical experiments, which are essentially limited in time. The experiment is set up (e.g. an electron is injected into an experimental chamber) then after an assigned lapse of time an observation is made which concludes that particular instance of the experiment. Outside this time-span there is no need to speak of a state at all. If, however, we do suppose that some state is definable after the observation, and that we are then quite free to choose some other observation to make, then the observational evidence forces us to assume that the state can change discontinuously and acausally every time it is observed.[7] The new state is determined by the outcome of the observation, an outcome which is quite unpredictable: the aim of quantum mechanical theories is to give ways of finding probability distributions for such outcomes.

This change is, from another point of view, described in the literature as the "collapse of the wave-packet".

3. THE CONFLICT

Two possibilities, which I shall describe in turn, now face us.

(a) The Two-world View[8]

We decide to limit the concept of "state" to the span of any particular experiment, making this a characteristic of descriptions of microphysics. Such a characterisation effectively divides the domain of science into two worlds: that in which clearly delimited experiments are carried out (e.g. particle physics) and that in which we participate in an ongoing state of affairs (e.g. astronomy). To the first is given a quantum description, while the second is described classically.

It is now crucial to realise that, while the general fields of the two studies may not be clearly distinct, the particular accounts which they give do not and cannot overlap (unless quantum mechanics be radically modified). This is because on any conventional quantum mechanical theory whatever, it is always possible to set up a quantum mechanical "state" which is a *superposition* of two other states which, on a macroscopic view (i.e. if they also had an interpretation in the other world), would be mutually exclusive. This, the heart of the conflict, has been given innumerable descriptions before, and I refer the reader who is unfamiliar with the idea to the very readable article by de Witt (1970).

A consequence of the inevitability of superpositions of "macroscopically opposed" states (that is states which, if they were part of the macroscopic description, would be exclusive) is the conclusion that the quantum and classical uses of the term "state" must differ: they are not talking about the same thing. Moreover, though the quantum description can be taken as an account of what happens in an atom, say, between its setting-up and its observation in any repetition of a given experiment, it cannot be taken as giving a complete account of any single instance of the experiment *as a whole* (including the final observation) unless an acausal link, analogous to "collapse", is postulated between the microscopic and macroscopic worlds. In itself, the quantum state provides information only about an ensemble of experiments, for whose results it provides a probability distribution. The classical description, on the other hand, refers in the first place to a single process; though it can then be applied to an ensemble of initial states to produce an ensemble of results with appropriate probabilities.

Thus the "maps" of the two regions interpenetrate and even touch but never overlap. The boundary of the quantum description can be joined approximately onto a part of the boundary of the classical one where probabilistic accounts of ensembles of experiments are concerned. It is known[9] that when the quantum description is carried to such large objects that the phase-information of the wave-function representing the state becomes small compared to the intrinsic thermodynamic uncertainty in the state, then, if the phase information is neglected, the quantum mechanical description can be translated into a classical probabilistic description. But where *individual processes* are concerned there can be no such translation.

One further consequence of this view needs noting: if the two worlds are distinct, the macroscopic world cannot be *made up of* microscopic entities. The table on which I am writing is not made up of atoms; an atom is only a state which is part of a description of some limited experiment to which I am quite external.

(b) A Unified View?

Unfortunately, this alternative does not really exist.[10] One would have to postulate that quantum mechanics and classical mechanics are in fact talking about the same things, but that the particular laws which govern large things are different from those which govern small ones. How, then, is the change-over to be made between the two? It might be possible to postulate some additional terms in the quantum mechanical equations of motion which only become significant on a large scale, but no way of doing this has found any general acceptance.

4. THE SCOPE OF OUR IGNORANCE

(a) Biology

This is the most striking example of a field in which there is a continuous transition from the microscopic to the macroscopic. Imagine, for instance, the processes involved in human vision: a single photon, which must be described quantum mechanically, falls on a rhodopsin molecule in a cone cell and is sufficient to trigger a nerve impulse which can affect the behaviour of the entire organism. Of course, the same is true of a Geiger-counter or of any particle detector: the difference is one of degree. In a particle-detector there is a fairly well-defined break between the microscopic processes and the macroscopic ones, to the extent that it would be unlikely to make any difference if one were to draw the dividing line in a different place. In the organism, on the other hand, there is every evidence of a finely tuned unity between the particle level and the molecular level, and between the molecular level and the structural. And it is precisely the links between these levels which are of interest biologically.

The two-worlds view must sever one of these links, regarding the rhodopsin molecule, for instance, as a quasi-classical observing apparatus when it receives the photon. A unified view, as far as one could envisage it now, could only postulate some unknown mechanism which inscrutably mediates the two worlds.

Although there is no evidence that the two-world view is false, a certain uneasiness attends it, particularly when one passes from biology to the psychological phenomenon of "free will". I would have liked to have catalogued this among areas of ignorance; but it is far from clear just what it is that we are ignorant of: despite millenia of attempts, no scientific language yet exists for turning one's vague feeling of mystery into sufficiently concrete questions. Whatever free will may be, it seems to have to do with organisms' unpredictability,[11] which may also be linked to the unpredictability that is peculiarly associated with the quantum domain (though there is no convincing proof of this).[12] At this point all one can do is to offer the speculative hope that a properly unified physics might dispel some of the obscurity which surrounds this most important topic.

(b) Astrophysics

So long as it is reasonably clear what is large and what is small (or, more precisely, where quantum interference effects are large and where they are negligible) then the two-worlds view will probably suffice and one could lay aside the desire for a unified physics as an outmoded relic of the nineteenth century. But in parts of astrophysics we seem obliged to speak of the large and the small in the same breath. Such is the case with theories in which the Universe has always expanded outwards and was arbitrarily small in the past. For the Universe of which we are now a part, must surely be classed among the large! The resolution of this paradox which seems to require us to apply quantum mechanics to the entire universe, is more than an academic exercise, since, on most current theories of cosmology, the fluctuations in density that must have been present at early times so as to form the galaxies and stars out of a homogeneous background cannot be explained classically, and appeal is usually made to the quantum realm of an even earlier epoch. Thus, without a resolution of the conflict, little progress can be made in explaining how everything around us came to be.

The problem has recently been presented in a time-reversed form. It is well known that most gravitational theories predict the formation of black holes. Now, if quantum processes are added onto a black hole background, then it has been shown by Hawking that the black hole must radiate energy, and in so doing radiate itself away to leave, in all probability, a "naked singularity": a point near which the length-scale of the gravitational field tends to zero and its strength tends to infinity (see Gibbons, 1976). Such an object can only be described quantum mechanically, and again we are faced with a conflict: the paradox of applying quantum theory to find the observable properties of an astrophysical object.

5. EPILOGUE

Many physicists and most biologists would, I surmise, argue that the problem I have described is (a) unimportant and (b) insoluble or inevitable, and it should therefore be ignored. Having given my counter-arguments to (a), I shall conclude with some hints as to (b). It may be that a position can be achieved (Clarke, 1974) which is less realist without relegating quantum mechanics to a computational algorithm. The result would be like the "guiding-wave" approach of the early interpreters of quantum mechanics,[13] with full reality granted to the observed states of the Universe, but a theoretical "reality" granted to a wave-mechanical model of a quantum universe which guided our observed states in a probabilistic way. This is, however, very far from a theory which will make knowledge possible in a region which seems condemned to ignorance.

NOTES

1. I use these terms in their scholastic sense, but generalize to psychological types; to the "Aristotelian" and "Platonist", if one wishes, though I would not take Aristotle as an eponym. The polarization is not unambiguous for philosophical theories (where do philosophical idealists stand, for instance?), but it is useful to apply it to their authors.
2. In the sense of Stapp (1972), not Bentham!
3. The sort of mathematical approach I have in mind is that of Mackey (1963), which, though considerably extended by later work, remains a clear exposition of the ideas. The philosophical literature on the subject is immense: the best single work I know of is d'Espagnat (1971); while a good popular discussion of the central conflict is the article by de Witt (1970).
4. This position has been taken, for instance, by Stapp (1972).

5. It might be argued the space-time itself provides a substratum, and that the "state of affairs" is simply the set of properties of space-time. This is more than a mere playing with words, if one goes on to postulate that the properties of space-time are adequately described by Einstein's vacuum equations; this is the Geometrodynamics of Wheeler (1962). I believe this is mistaken, since Einstein's equations are essentially macroscopic, both as regards their conceptual background and their verification.

6. For the sake of simplicity I have used the Newtonian presupposition that space and time can be separated. Incorporation of relativistic ideas complicates the situation and makes the conflict a lot worse, since it becomes difficult to delineate the "beginning" and "end" of an experiment in many cases. A full discussion is in Clarke (1976).

7. There are theories, such as the Bohm and Bub (1966) hidden-variable theory in which causes are postulated for a quick but continuous change. But theories like this require one to "build-in" to the particle a series of hidden variables which will account for each of the experiments which it will subsequently encounter, as if all particles were created with a joint foreknowledge of what these were to be and a programme for what outcomes to produce. This is not causality in the scientific sense, but more a pre-established harmony. The alternative position, of collapse, is stated most uncompromisingly by von Neumann (1955).

8. It is only fair to add that this sort of view can be developed so as to become much more acceptable than the crude version that I have given in terms of maps. When thus developed, it becomes essentially the "complementarity" approach of Heisenberg (1974). While this may be the best that one can do, the defects to which I draw attention still remain.

9. This was most decisively demonstrated by Daneri, Loinger and Prosperi (1962, 1966), whose philosophical importance was discussed by Rosenfeld (1965). It will be seen that Rosenfeld's discussion is totally linked to the domain in which clearly delimited experiments are performed by outside observers. Further development has been carried out by Prigogine (1969), George (1972) and coworkers.

10. Which is not to deny that there are many attempts, some of which are successful, but all either *ad hoc* and contrived (including my own, Clarke (1971)), or else requiring such a radical revision of everything that they are a long way from physical utility. As a random selection, I would mention Bastin (1966, 1971) and Bohm and Vigier (1954). This latter paper has received negligible support because of the rather grossly classical framework in which it was presented; it has many formal features, however, which could be successful. A thorough-going attempt to construct a unified "atomism" has been given by H. P. Noyes (work to appear). A successful approach would have to start from foundations that differed radically from both classical and quantum physics as we know them, and then derive both as limiting cases. "Glueing together" classical and quantum physics is probably doomed to failure.

11. Without attempting to solve the problem of free will in a footnote, some expansion should perhaps be made here. First, it is not obvious that free will and determinism are incompatible. This has been extensively explored in the case of the "determinism" associated with predestination, where even Augustine (often erroneously thought to be a denigrator of free will) allowed our free choices to be part of the pattern of causation without their being thereby less free (*City of God*, Books XI, XII), and this view was explicitly espoused by Abelard. However, in an obviously deterministic Universe free will appears somehow gratuitous: one is forced to regard it as an epiphenomenon, and so as somehow fraudulent. A measure of indeterminism, whether it comes from thermodynamics or quantum theory, leaves the door wider open for free will, without thereby saying what it might be or what it does. And this happens more clearly with quantum theory, which, on most interpretations, has a kind of irreducible indeterminism built into it (Margenau, 1965). But then one must go on to say what free will is, if it is neither a quantum mechanical "hidden variable" (which, even if possible – see note 7 – would reduce everything to causality again), nor another name for mere randomness. I have attempted the first steps (Clarke, 1975) in constructing an intermediate status of "spontaneity", which exploits the mathematics of infinite vector spaces to allow the existence of a non-temporal causative factor which moulds the progress of events in time so that they acquire short-term significance but long-term randomness.

12. It is significant that very many critical processes in biology are associated with the cleavage of hydrogen bonds, whose energy lies in the range of thermodynamic, not quantum, energy fluctuations. The possible role of quantum processes could be in setting up co-operative effects that introduced a coordinated uncertainty in the timing of cellular events that could not be achieved by purely thermodynamic means.

13. Compare, for instance, the description of Born's views given in Jammer (1966).

REFERENCES

Ballentine, L. E. *et al.* (1971) *Physics Today*, April 1971.

Bastin, E. W. (1966) *Studia Philosophica Gandensia* 4, 77.

Bastin, E. W. (1971) *Quantum Theory and Beyond*, Cambridge U.P., pp. 213-26.

Bohm, D. and Bub, J. (1966) *Rev. Mod. Phys.* 38, 453-69.

Bohm, D. and Vigier, J. P. (1954) *Phys. Rev.* 96, 208-16.

Clarke, C. J. S. (1971) *Int. J. Theoret. Phys.* 8, 231-5.

Clarke, C. J. S. (1974) *Phil. Sci.* 41, 317-32.

Clarke, C. J. S. (1974) "Eternal Life", *Theoria to Theory* 10, 181-9.

Clarke, C. J. S. (1977) "Uncertain Cosmology", in *The Uncertainty Principle*, Ed. S. Chisick, Wiley, London/New York.

Daneri, A., Loinger, A. and Prosperi, G. M. (1962) *Nucl. Phys.* 33, 297.

Daneri, A., Loinger, A. and Prosperi, G. M. (1966) *Nuov. Cim.* (B) 44, 119-28.

d'Espagnat, B. (1971) *The Conceptual Foundations of Quantum Mechanics*, Benjamin, Menlo Park.

de Witt, B. S. (1970) *Physics Today*, Sept. 1970, p. 30.

George, C., Prigogine, J. and Rosenfeld, L. (1972) *Kong. Danske. Vidensk. Selsk. Mat.-Fys. Medd.* 38, 3-44.

Gibbons, G. (1976) *New Scientist*, 69, 54-56.

Heisenberg, W. (1974) *Double Dialogue, Theoria to Theory*, 8, 11-34.

Jammer, M. (1966) *The Conceptual Development of Quantum Mechanics*, McGraw-Hill, New York.

Mackey, G. W. (1963) *The Mathematical Foundations of Quantum Mechanics*, Benjamin, New York/Amsterdam.

Margenau, H. (1965) *Phil. Sci.* 30, 1.

Prigogine, I., George, C. and Henin, F. (1969) *Physica* 45, 418.

Rosenfeld, L. (1965) *Suppl. Prog. Theoret. Phys;* Extra number, p. 222.

Stapp, L. P. (1972) *Am. J. Phys.* 40, 11.

von Neumann, J. (1955) *Mathematical Foundations of Quantum Mechanics*, Princeton U.P.

Wheeler, J. A. (1962) *Geometrodynamics*, Academic Press, New York.

Ted Bastin

Strongly influenced as a student by Eddington's vision of the nature of the quantum. Studied at Queen Mary College, London, and the University of Cambridge. Formerly Research Fellow of King's College, Cambridge.

Research interests include foundations of quantum theory and the limitations of metrical concepts of the very large and the very small.

A CLASH OF PARADIGMS IN PHYSICS

To be asked to write on areas of our scientific ignorance is ɩo be given an unusual and valuable invitation. Usually a scientist is asked to write about what he can demonstrate in detail, even though in deciding his orientation he will presumably have taken into account what he sees the important area of ignorance to be. With his title "The Art of the Soluble" Sir Peter Medawar has indicated a very important aspect of science which goes some way to justifying the usual approach. You do what you can do and you build on what has been done: you do not usually confront ignorance head-on.

It follows, however, that the very vision of the world that the science of a particular day gives you is heavily conditioned by what problems one can reasonably expect will turn out to be soluble given the existing methods and techniques (theoretical and mathematical), as well as experimental, that are available in that day. The areas of ignorance which are not penetrated, or at least not illuminated, by the existing methods and techniques tend to be overlooked. Sometimes it is held that they cannot exist, and we can surely expect better of scientists than that!

Still, however much one expects that scientific innovators should be sensitive to areas of ignorance, one would not expect them to *wallow* in ignorance. Some of the writers, for example, who advocate vitalism – the ones whom we may usefully call the "thus-far-and-no-further vitalists" – seem to rejoice in the possibility that there may be areas which are for some reason forever closed to rational investigations. Such a position seems really contrary to the scientific spirit. So we may well ask, under what circumstances could it be reasonable to allow one's scientific endeavour to be guided by a vague but overriding conviction that the existing approaches are fundamentally wrong even though no technique or organised way of thinking exists which could conceivably provide a better alternative? What I am seeking, therefore, is a way in which a survey of our ignorance might serve a scientifically creative end (*pace* Wittgenstein's aphorism "the limits of my language are the limits of my world" which in general sums up so much about the scientific vision).

I think that the current situation in the foundations of physics is one where a survey of ignorance can be valuable. I shall argue that there are two paradigms[1] discernible in this current situation which are incompatible. One, which I call "the classical paradigm", is so familiar in its application that it practically constitutes physicists' thinking as we have it at the moment. The other – which I shall call "the sequential paradigm" – has been forced into existence by our experimental knowledge but has no background of thinking out of which it naturally emerges. The latter is the one we need in our present situation; the former is the one which alone we can think with. The resulting conceptual confusion is what I shall mainly explore as my contribution to this "encyclopaedia of ignorance". To wind up, I shall probe into areas of even darker ignorance by suggesting that if we face this confusion we may find the strange phenomena of psychokinesis less perverse.

The "classical paradigm" has its basis in the idea of a continuous background of space and time, which are imagined as perfectly smooth, perfectly homogeneous, infinitely divisible continua, mathematically modelled by the continuum of all real numbers as it was finally formalized by Dedekind and Cantor between 1870 and 1880. Physical entities are located within these continua of space and time. The basic entities are particles – idealised as single points whose position in space changes continuously and smoothly with time – and fields – distributed through space and at each point of space having a certain intensity which again varies smoothly both with time and with changes in the special point at which the field is considered. The elegance of the classical paradigm as it appears in, say, Thomson and Tait's *Treatise on Natural Philosophy* (1887) compelled a feeling of universality; one was in possession of the way to describe reality direct. Indeed the classical paradigm has become for the physicist a mathematical elaboration of common sense and the automatic vehicle of his thought. And the quantum theory has produced a formalism which has strikingly different implications from what is implied by the classical paradigm without providing the new way of thinking that one would demand of a true change of paradigm. Indeed Niels Bohr, the profoundest thinker among the originators of the quantum theory, knew that there was a reality other than that described by the classical paradigm, and yet thought it inconceivable that we could learn to think about the world in terms other than those provided by the classical paradigm. He is reported to have said, "you might as well say that you and I are not sitting here drinking tea" (which is what they were doing) when this possibility was put to him. Bohr thought that one could consider phenomena under different aspects using the spectacles constituted by the classical paradigm, and that the limitations inherent in that form of description and understanding would show themselves as incompatibilities between these aspects as they were described. Bohr's complementarity philosophy for the foundations of quantum theory was based on the existence of these incompatibilities, and he was happy with this philosophy with its neo-Kantian flavour. Not merely did he think it impossible to change the spectacles; he did not want to.

Physicists in general have not the drive for consistency of thought of Bohr, and do at present think about the fundamental particles of physics in ways that I believe to be incompatible with the classical paradigm, while continuing to assume the general validity of the familiar concepts — field, momentum, mass, force and so on — that derive their currency within the classical paradigm.

If I could give a clear description of the new ways of thinking which I see to be intermittently emerging in the study of the fundamental particles — and to which I have referred as the "sequential paradigm" — I should not be writing this essay using it as an example of our ignorance. However, we get a clue from the fact that our direct knowledge of the fundamental particles is derived entirely from the processes of disintegration and association which they undergo. These we imagine to take place in sequences which we, as experimenters, interrupt. We call the point of our experimental interruption an *observation*. From such an observation or from a very small number of coupled observations we make inferences about the progressively more remote events in the sequences, but these inferences become rapidly more conjectural and dependent upon theoretical extrapolation as they become removed successive stages from the point at which the experimenter's interruption takes place.

Most of the time one thinks of a background to the experimentation on particles which is permanent, objective, and accessible to observations of the kind that we are familiar with from everyday, large-scale physics. A background that fits the classical paradigm, in fact, and into which the sequences of atomic events fit. Indeed one usually thinks of the macroscopic world as being *made up of* or *constituted of* the entities which take part in these sequences of events.

I conformed to this way of thinking myself in writing just now of the "interruption" by the experimenter when he makes an observation. Strictly speaking — within the way of thinking that I envisage — one knows only about such a background of events which are interrupted, through the information obtained from just such interruptions, and therefore the assumption that there is something going on in the background which can be described independently of the interruptions, is one that has to be demonstrated to be valid whenever it is used. What seems to be needed is a way of thinking in which apologies like these have not continually to be made for our inappropriate upbringing. Such a way of thinking would automatically steer us along lines which are realistic because they correspond to the operational situation we are actually in, by forcing us to see the world form the aspect of a point in a branching sequence of atomic events. This would be our sequential paradigm.

The first clear intimations to the world of physics of the facts which I claim to demand a new "sequential" paradigm resulted in the growth of the quantum theory. The quantum theory was elaborated over quite a long period, to take into account a certain class of experimental facts – namely, those facts which forced on our attention *discrete* attributes of the physical world which cannot be incorporated within an essentially continuous classical theory. It is reasonable to ask how far quantum theory has succeeded in the task of extending or replacing the classical continuum picture so as to incorporate these attributes.

The early forms of the quantum theory never attempted to *explain* discreteness in the sense that they could be said to have incorporated both the discrete and the continuous within one theoretical structure. They simply imposed discreteness as a mathematical constraint upon the ranges of values that were allowed to be interpreted as the results of measurements of certain physical quantities. As the quantum theory developed, however, physicists' attitudes to the problem of explaining the discrete values changed, and it came to be accepted that the modern (1926-1930) form of the theory was an intellectual structure within which discrete and continuous quantities could cohere.

If this were true, my concern with a sequential paradigm could be no more than a search for a prop to the imagination in solving problems for which a complete solution already existed in principle. But is it true? It is not now as easy as it was a quarter of a century ago to ignore the disquiet which many physicists have continued to voice about basic inconsistencies in the foundations of the quantum theory. This disquiet has been focused largely on what is called "the measurement problem" though this expression conceals muddled thought. To speak as though there were this one remaining problem to solve before quantum theory could be made properly consistent is a bit like saying that the project of finding an overland route to Australia was waiting only for the solution of a last remaining problem – the "closure problem" they might call it – constituted by the closed perimeter of that continent.

Physicists have felt justified in ignoring logical flaws simply because of the vast successes of the quantum theory. How could a theory be wrong which works so well and with such precision in so many detailed problems? Answer does indeed have to be found to this question if one proposes to question the universal applicability of the quantum theory, and the one I propose is to say that the successes of the quantum theory are without exception *combinatorial* in character, and therefore their experimental accuracy – however great – does not argue for the thesis that the quantum theoretical formalism constitutes a new dynamics to replace or generalise the classical one. I believe that my point of view can be justified though even to demonstrate its validity as a logical possibility would take great space, while to work it out in detail as an established point of view in theoretical physics would be to take on several major investigations. All I shall do to convey what I mean by "combinatorial" is to draw an analogy. Consider the proposition that the back wheel of a particular bicycle rotates exactly 2.4 times every time the pedals are turned once. I say that this proposition is *combinatorial* because it can be deduced from the observation that the numbers of teeth on the two sprockets are in the ratio 5:12. It is also true that this combinatorial relationship exists in a physical mechanism whose dynamics we understand in a fair degree of detail: such things as the strengths of the various parts of the structure of the frame, and the behaviour of the material of the tyres as well as the dynamical relationships of the parts – gyroscopic effects and so on – which contribute to the balance of the bicycle are all described to a high degree of approximation by classical mechanics. The value for the relative rates of rotation of the wheels, however, is not a part of this corpus of knowledge, even though one could measure it to a high degree of approximation experimentally. For no one would suppose that the high accuracy with which such experiments might give the value 5:12 would constitute an argument for the correctness of any of the branches of classical physics which had contributed to that corpus of knowledge.

The sequential paradigm that I have been arguing the need for must be combinatorial in character. The events it describes must happen as do the moves in a game of chess, and as with the moves in a game of chess the shape of the game depends upon the different possibilities of play that are allowed — again all combinatorial relationships.

The form of the quantum theory as we have it affords many clues to the way to go to work to set up a combinatorial form of that theory. For example, the well-known spin vectors — being two-valued — fit far more naturally into a combinatorial picture than they do into any conventional dynamics. The fact that physical quantities like the spin vectors (the old "quantum numbers", in fact, and their more modern counterparts) seem to need to be fitted into a combinatorial picture takes us only half-way to a sequential paradigm and beyond this half-way point our way seems blocked. There is good reason for this block.

A great deal of effort (one might say the central effort) of the quantum physicists of the past and present generations has gone into a search for a combinatorial scheme from which the quantum numbers and discrete attributes in general of the wide range of elementary particles that are now known could be deduced with as much consistency as, for example, was obtained for the energy levels in the low energy quantum theory. It is very difficult to say whether this search has been successful in any significant way. It is universally agreed that definitive success has not been achieved, in spite of the length and intensity of the search.

One therefore, naturally, looks at the raw materials that have been used for such combinatorial schemes — the origin of the bare numerical relationships from which they could be built up. One then discovers that in all the current approaches these numerical relationships have been sought in an essentially geometrical setting. The mathematical techniques used have often been very abstract but they have remained abstractions from what one could call, in the widest sense, geometry. Thus, for example, a simple combinatorial relationship could be derived from the connectivity of the Möbius strip. If a strip of paper is bent round and stuck onto itself like a link of an old-fashioned paper chain, and then cut with scissors along its middle line, then it will fall into two rings. If, however, one end of the strip is twisted through 180° before being stuck, then the scissors will produce *one* twisted link of twice the length. The half rotation has been responsible for a change from one object to two, and the whole phenomenon might well be taken for a model of the origin of how combinatorial relationship might be linked to something happening in the world, and hence the basis for the numerical value of, perhaps, a mass ratio in a fundamental particle. The process, however, which is imagined to be physical in character — namely the twisting and cutting of the strip — is essentially geometrical (or, more accurately, topological).

There is a small number of theorists who acknowledge the profound difficulties in assimilating any theory of the quantum phenomena to a continuum theory and who are convinced that the right course of action is to build the continua of space and of time from discrete processes of some kind. Even they, however, seem to fall back upon a geometrical origin for the combinatorial relationships that shall characterize the discrete processes, in spite of the obvious danger of circular argument that is entailed. In the "twistor" theory of Penrose and his associates, for example, the attributes of the different types of particle are sought by considering the possible associations and combinations of certain abstract group structures. One would suppose that the amount of freedom and power given by this approach would be exploited by using the very abstractness of the groups to import new kinds of connectivity and to get away from the familiar physical picture. What we find, on the contrary, is that the groups used are the familiar groups that describe the translations and rotations of bodies in 3- and 4-space. There seems a failure of the imagination at this crucial point.

Of course, I have already supplied my own answer to the question "how should the situation be exploited: what is the imaginative leap that is needed?" in talking about a sequential paradigm. We need a mathematics in which the dynamics of objects moving in space is replaced by a dynamics of branching sequences of events with the decisions at each branch point of which branch to take, itself contained in the system. In fact the dynamics is that of a computer programme.

It is not that the details of computer techniques, let alone computer engineering, are particularly relevant to the situation. It is rather that computer people work all day with relationships which are those required for a sequential paradigm whereas physicists work with ideas which are foreign to it. Not surprisingly the appropriate expertise is to be sought among the former. There one finds certain broad principles which are common to computer practice even though they have not yet been made the subject of an elegant formal mathematical system. They do not depend upon the particular ways the computers are designed, and certainly not upon what materials they are made of. These principles are the bare bones of a non-existent logic of sequential relationships.

They are:

1. That there exists a set of initial elements which have a structure represented most simply by an ordered set of 0s and 1, which fill positions in a string.
2. That these elements have labels or names expressed by numbers, which enable the elements to be retrieved. This provision is equivalent to providing a "store" or "memory".
3. That there exists a processing unit to which elements can be brought in pairs and in which logical operations can be performed.
4. There exists a process for creating new elements (and destroying them).
5. Each process or operation contains or generates instructions for bringing a "next" process into play, so that the working of the system proceeds automatically.

If we now consider the elements of our working system as corresponding to the "points" of the physicists' space, we see the strangeness of the new ideas. Instead of the freedom which the physicist (and the classical mathematician) assume as their right, to imagine themselves selecting one point or other as they wish, for consideration, secure in the knowledge that there is an intuitively obvious spatial relationship to guide them from one point to the next, these relationships have to be constructed before they have a meaning.

My proposal is not entirely without precedent. As a programme it has recently been suggested by Finkelstein.[2] With John Anson, C. W. Kilmister, Pierre Noyes and A. F. Parker-Rhodes and others I have been working on such ideas for some time and have produced principles relating especially to point (4) above in which the relative scales of the basic physical forces can be deduced from a very fundamental process of generating sets of points of increasing size by incorporating the operations on the sets into the sets. Noyes[4] is working from the point of view of high-energy theoretical practice to get an approach which can incorporate such principles. Looking back into history, I see the nearest thing to an origin for such ideas in Whitehead's efforts to consider spatial relationships as logical relationships.[3]

At this point in my argument I bring in a quite new kind of evidence — from psychokinesis. It happens that I have myself had extensive and variegated first-hand experience of experimentation with two well-known subjects who are able to influence physical objects in a paranormal way (that is to say, to exercise psychokinesis). These are Uri Geller and Suzanne Padfield. I have seen odd examples of psychokinesis produced by several other subjects. I mention the personal origin of my evidence only because the subject of psychokinesis has been much in the public eye and opinions are at the moment balanced between acceptance of it as a real phenomenon and rejection of it as too implausible to be considered. Little can be done by one person's experimental evidence to alter this balance (though evidence does have a cumulative effect) because the credibility of witnesses and experimenters is being called in question — and this is the case whichever side their evidence supports. Moreover, physicists with a lot of first-hand experience of psychokinetic effects are still not common.

Psychokinetic effects show an effect of "thought forms" directly on physical matter. By this I mean that the way a subject thinks about an object strongly influences what happens to the object. Physical effects which are influenced by thought forms are quite unlike the effects of physical forces of the familiar kind, and this dissimilarity is crucial.

It is very difficult to convey a sense of what I mean in a short space, but I can give some idea of the strangeness of what goes on if I point out that the effect a subject will have will depend upon the way the subject divides up the matter with which he is surrounded in his mind into "objects" with individuality. If a subject succeeds in having a psycho-kinetic effect at a distance on a spoon, then that spoon, which may be one among a heap of assorted objects of variegated materials, will be singled out: perhaps bent, perhaps apported,[†] perhaps both.

No ordinary physical force will do anything like this. A powerful neutron beam may have very odd effects on a heap of objects, but it will not know that one piece of metal is part of, say, a watch — and an adjacent piece part of the tray on which the objects are placed. Until one has faced up to this central fact about psychokinesis one has not faced up to psychokinesis. Many scientists who have set out with a serious intent of studying psychokinesis have been so put off by the obstinate intrusion of what I have called "thought forms" of this kind that — contrary to their own expectations — they have joined the ranks of those who did not want to know.

I do not believe myself that the fact that one has to come to terms with "thought forms" which have the power over matter of physical forces means one has to abandon rational inquiry. One does have to look for a minimal change in basic theory though, which will incorporate the thought forms without trivializing them. This is where my discussion of psychokinesis connects with the sequential paradigm, for I believe that the "minimal change" for which we have to look is just the inclusion of memory into physics which we have seen the sequential paradigm to demand.

I described earlier how violent a change the sequential paradigm would demand in what we take as "common sense" or "just what is to be expected". The classical laws of mechanics would no longer be the point from which investigations begin but would become the point at which some special deductive schemes with highly restrictive conditions might ultimately finish. Getting there at all would be a struggle, and the main difficulty would be to establish the persistence of macroscopic objects and the massive impression of objectivity they provide. However, those are the *difficulties* of the new theory: let us turn to what it would cope with rather naturally as soon as our understanding of the rudimentary interaction processes had led the way to something a bit more complex.

It seems plausible that the example we know best of a memory system — namely the human

---

[†] An object is said to be "apported" when it is transferred from one place to another by psychic means.

being — may play the same part in our sequential physics as the complicated classical physical object plays in the old paradigm: namely, that of the measuring instrument with which we get our direct information about the elementary processes. Of course, we should not expect to get a deductive build-up from the elementary sequences to the complexity of a human being any more than we ever get a deductive build-up from the atomic building bricks of classical (let alone quantum) physics to a thing like an armchair or a steam engine (the argument that these deductions are there "in principle" is just an act of faith which we can appeal to with our complex system no less — and no more).

The different position we would be in with our sequential paradigm would therefore be that whereas we should find it very difficult to give an account of a stable objective inorganic universe, we should find it not such a great step as to defeat the imagination to think of specific physical objects which incorporate histories which make them accessible to particular sensitives — persons who are the good psychokinetic subjects.

When this very unpredictable and special rapport exists between these structures (namely the history of the object and an atomically specified memory trace in the human subject) then changes of the kind we call psychokinesis, and involving thought forms produced by the subject in his efforts to elicit these memory traces, will take place.

Let us look at the changed situation another way. Our new paradigm frees us from the preconceptions of spatial and temporal and what it gives us instead to play about with un-selfconsciously (as we may put it) is *similarity of pattern*. In so far as two physical entities have a pattern in common which is specified at the microscopic or quantum level, there ceases to be any problem of how they interact. They do so automatically; for to the extent that they possess similarity of pattern they are to that extent the same entity, even though, as we should normally consider them, they are separated in space or in time.

If now, we suppose that the human being — presumably through the operation of his brain tissue — is endowed with extreme sensitivity to the patterns which exist in physical objects as a result of the particular circumstances and situations through which they have gone in their past, then we have no need of any further hypothesis to explain how interaction takes place in spite of the absence of mechanical connection. In the case where one entity is a physical object, therefore, and the other a human being, one should imagine that the human being is capable of reproducing a part of the pattern which is constituted by a temporal sequence of events in the history of the object. If we want to think temporally we can imagine the brain of the person running along the paths into the past of the object. But we need not do so; we can equally well think of a whole path of this kind run together to form a more complex pattern which is directly comparable to one in the brain. In the context of this speculation it is very relevant that sensitives are usually very much more easily able to affect objects which have played an interesting part in the lives of some other person or persons (like watches and ornaments) than they are able to affect mass-produced artefacts.

My account of psychokinesis is no more than a sketch. I have not even suggested how, for example, a physical object might include in its structure coded reference to the events in which it took part, and there are several equally glaring gaps. On the other hand, there are gaps in our understanding of the relation of brain tissue to the workings of the imagination which amount to almost total ignorance anyway. Which takes me back to my beginning: familiarity may make us see a reasonable coherence where in fact there are great areas of ignorance while denying any coherence to unfamiliar ideas which may be no worse in their incoherence. Therefore to juxtapose an unfamiliar picture with the familiar may be a very good way to construct our encyclopaedia of ignorance.

NOTES

1. "Paradigm" is the word Thomas Kuhn has associated with the view that scientific change characteristically takes place by a revolution in ways of thinking rather than by continuous change. The paradigm is the essential kernel of the way of thinking. I subscribe to the revolutionary view and use the term "paradigm" in the general sense that I have just defined, and not necessarily in any of Kuhn's detailed senses.

2. See, for example, David Finkelstein, "Space-time Code IV". *Phys. Rev.* D.9,8 (15 April 1974). Also, contribution to *Conference on Quantum Theory and Structures of Space and Time, Feldafing*, July 1974.

3. Pierre Noyes, S.L.A.C., Stanford University, California. The only account published as yet is "A Democritean Phenomenology for Quantum Scattering Theory". *Found. Phys.* 6, 83 (1976).

4. A. N. Whitehead, "On Mathematical Concepts of the Material World". *Phil. Trans. Roy. Soc.* (1906). Also reprinted in *A. N. Whitehead, an Anthology*, edited by Northrop and Gross, Cambridge, 1953.

Sir Alan Cottrell, F.R.S.

Master of Jesus College, Cambridge.

Deputy Chief Scientific Adviser to H.M. Government 1968-71 and Chief Scientific Adviser to H.M. Government 1971-4.

Awarded numerous honorary degrees and prizes including the Harvey Prize, Technion, Israel, 1974, and the Rumford Medal of the Royal Society, London, 1974.

Main interest has been in the atomic theory of the properties of matter, especially those of metals, with emphasis on the theory of the strength, ductility and brittleness of steel, and also on problems of nuclear radiation in solids in connection with the development of nuclear power. More recently interested in the roles of science and technology in national affairs and also in industrial policy. Another recent interest is in the presentation of science to the general public.

EMERGENT PROPERTIES OF COMPLEX SYSTEMS

WHOLES AND PARTS

How does quantity become quality? How do the distinctive properties of bulk matter emerge out of those of its constituent particles? How do the characteristic properties of, say, a plasma, a superconductor, or an insulator, emerge from those of the charged particles in them; or those of diamond or graphite emerge from those of the carbon atom; or the self-reproduction of DNA or the enzymatic action of protein, from organic molecules; or the self-awareness of mind, from neurons; or the significance of a newspaper picture, from a set of dots? Such questions bring up the general problem of the origin and nature of *emergent properties*, i.e. properties of a whole system not possessed by its parts.

Even by setting out the problem in this way we have already taken up a definite point of view, which is the orthodox scientific one that wholes are in principle — if not yet always in practice — entirely explicable in terms of their parts. In the light of so many triumphs today in the application of atomic theory and quantum mechanics to the understanding of the structure and properties of bulk matter, no great courage or originality is demanded by such an attitude, certainly within the physical sciences and perhaps also in molecular biology. But when we come to the behavioural, psychological and social sciences, too little is understood, yet, for there to exist a scientifically objective basis for rejecting the old vitalist belief that some of the emergent properties of life cannot be completely reduced to physics and chemistry. However, irrespective of whether one believes in vitalism or reductionism, it remains a sound research tactic to proceed on the working assumption that all wholes are in principle understandable entirely in terms of their parts, since this has been so consistently successful in the natural sciences and since it may be the means of bringing science right up to the edge of the supposed gap between the material and mental worlds, a gap which may then be seen to be either illusory or profound.

In a sense, most of the ordinary properties of bulk matter are emergent properties since the only fundamental properties in physical systems are the kinematic properties of elementary particles. Exact science begins with the laws of *motion* and motion is kinematics. Even forces have no fundamental significance, being only auxiliary concepts introduced because they enable the correlations in the motions of two different particles to be neatly epitomised. All other features of the physical world, as we experience them, derive from the kinematic properties of elementary particles *and from our imperfect knowledge of them*. When we stand in front of a fire, the movements of electrons in the flames bring about corresponding movements of other electrons in our eyes and skin; and we see light and feel heat. From the movements of molecules in a gas emerge the properties of pressure, conduction and convection. The frictional properties of matter or, more generally, the properties associated with thermodynamic irreversibility, do not exist at the level of elementary particles. They arise entirely from the fact that we can recognise and describe *certain* overall kinematic states of large groups of particles — those in which the individuals share a common motion or some other uniform kinematic feature — *more simply* than we can recognise and describe any members of the much larger class of disordered states which lack such distinctive features. We thus meet here a most important feature of complex systems: *some new properties of matter emerging out of our ignorance of the individual motions of its constituent particles*. Moreover, it is not merely a lack of knowledge but also a lack of *interest*. For, if we had a science of bulk matter which told us just where every particle is and how it is moving (which is, of course, quantum-mechanically impossible), delivered as a vast computer print-out of billions of positions and velocities, we would then *know* everything about the piece of matter in question but *understand* nothing about it. It is not merely the imperfections of our senses which cause new bulk properties of matter to emerge from large assemblies of particles; it is also the nature of our minds, which crave understanding and seek only such general knowledge as is necessary for that understanding. It is our own subjective qualities, even when we are practising rigorously as natural scientists, that open a Pandora's box of emergent properties of complex systems.

ORDER, DISORDER AND ORGANISATION

Since a complex system contains many particles which constrain one another's positions and motions — and thereby develop a *structure* between themselves — we have to consider structure as well as properties. For systems such as crystals it has been useful to distinguish between *simple* and *complex* distributions of the particles in a structure. Consider from this point of view a large, regular checkerboard of $2N$ squares, each of which is given a single nought or cross selected from N noughts and N crosses. Regard the board as two interpenetrating square grids, labelled A and B respectively. Then an example of a simple distribution is that described by the statement "all the noughts are on grid A, and the crosses on grid B". The geometrical and conceptual simplicity of this distribution are displayed by the fact that it is *completely* describable in a sentence of a *few* words. By contrast, a complex distribution lacks this special feature and leaves us with no option but to describe it in a long cataloguing sentence which separately reports the particular nought or cross state of each individual square.

A *simple* distribution is, of course, an *ordered* one. In physical systems such as solids, liquids and gases it has been quite sound to infer also the opposite to this, i.e. that a *complex* distribution is a *disordered* one. But this is not generally true in biological systems, because here some complex distributions are specifically produced from detailed instructions and have spectacularly different properties from the usual type of disordered distribution.

Such systems, i.e. DNA and protein molecules, are not ordered, at least in the sense by which they might be regarded as structurally simple. They require long sentences for their description and, considered purely structurally, appear to be disordered. But they are not disordered, either. They have unique properties which are critically dependent on their precise structures. We may call them *organised* systems and regard them as a second and distinct class of complex systems which are produced by biological action.[1] The contrast between *order* and *organisation* is made clear by comparing, say, a crystal with an amoeba. The first is highly ordered. The amoeba is not at all well ordered, since it consists of a shapeless bag containing a sticky fluid in which float irregularly shaped long-chain molecules. But it is organised to a sophisticated extent that leaves the crystal far behind and it has some spectacular emergent properties: it can feed and keep itself alive, adapt to different circumstances, and make replicas. Another example of an organised system is the haemoglobin molecule. Its special ability to capture or release oxygen depends on a configurational change of the whole molecule, so that the complexity of the structure is essential, for the special chemical properties of the molecule, and in fact is specifically organised to provide just those properties.

PHYSICO-CHEMICAL ANALOGUES OF BIOLOGICAL PROCESSES

The link between purposive organisms, such as the amoeba, and ordered crystals, is provided by molecular biology. The properties of the DNA molecule, though spectacular, are nevertheless still recognisably "physico-chemical". Its ability to reproduce itself — by each of the arms of its double spiral, when separated from the other, serving as a template for the construction of a replica of the other — is structurally an extreme elaboration of the ability of one crystal to seed others, in a saturated solution, and is kinetically an example of the auto-catalysis which occurs in many chemical reactions. An important difference, however, is that DNA can exist in many different distributions, according to the sequence of nucleic acid units along its arms, and these differences profoundly influence the relative abilities of the various forms of the molecule to compete for survival by reproduction. Of course, some crystals are better fitted for competitive growth than others, for example by the presence in them of dislocations which provide growth spirals on their surfaces, but this is not a useful analogy. Moreover, by means of mutations, which alter the sequences of nucleic acids along the arms, new varieties of DNA emerge from the general population, some of which may be even more competitive for survival than their predecessors. Out of this emerges the great biological property of self-improvement by evolution.

As well as seeking physico-chemical foundations for biological processes, people have constructed mathematical models of these processes, by generalising the kinds of expressions familiar in theories of chemical kinetics and of quantum-mechanical transitions. One of the most interesting of these is due to Eigen,[2] who constructed equations for the concentrations of nucleic acid polymers for given rates of multiplication (auto-catalytic), mutation and decomposition, subject to an overall conservation condition which leads to competition and hence "natural selection" between the different molecular varieties. In this way he was able to derive solutions which display the basic features of evolutionary biology. In this mathematical model the mutations occur indeterministically (stochastically) but, once started, the populations of new strains develop deterministically. Eigen's system is thus *unstable* against small spontaneous fluctuations (i.e. mutations) and it is from this instability that its ability to evolve "biologically" emerges.

EMERGENCE OF MACROSCOPIC STRUCTURES AND PROPERTIES

A more general analysis of the emergence of macroscopic structures and properties of unstable systems has been made by Glansdorff and Prigogine.[3] Living organisms are examples of *open systems*, which receive highly ordered energy (e.g. sunlight) from their surroundings and use some of this order to maintain and construct themselves. In this respect they are like heat engines and heat pumps which also concentrate energy by operating on an energy flow. There is a well-developed subject of irreversible thermodynamics which deals with energy flows and derives relations such as Ohm's law, Fourier's law, etc., but this is limited to systems very near to thermodynamic equilibrium, systems that attain stable steady states by attempting to relax to equilibrium in the face of the small and continually perturbing effect of their imposed non-equilibrium boundary conditions (e.g. different temperatures across a heat exchanger).

Glansdorff and Prigogine have shown, however, that when the deviation from equilibrium and the ensuing flows go beyond a certain threshold, the system is often no longer stable in a steady state. As a manifestation of this instability a *macroscopic structure*, such as a convection cell in a heated fluid, may appear spontaneously in the system. Below this threshold, the energy all goes into the individual thermal motions of the particles, but above it some is channelled into energy of macroscopic structures and stream patterns. The most interesting examples occur in chemically reactive systems, in which various reactants move through the system by diffusion, as well as being created or annihilated by chemical reaction. Depending on the chemical conditions, in such systems above a certain size, there can emerge spontaneously rhythmic oscillations in the concentrations of the reactants (analogous to the predator-prey cycles in ecological systems, or the depression-boom cycles in industrial economies); and also in some cases there emerge spatial patterns which break the originally homogeneous symmetry of the system. Such spontaneously formed oscillations and macroscopic patterns – which exemplify Herbert Spencer's idea of the "instability of the homogeneous" as the foundation of natural evolution – have been studied experimentally by Zhabotinsky and Zaikin[4] and by Winfree,[5] and were earlier predicted theoretically in a remarkable paper by Turing.[6] The instability of the homogeneous may also be significant in the theory of the early evolution of the Universe.

These extensions of classical thermodynamics and kinetic theory to conditions far from equilibrium are an important development which is enabling a new bridge to be built from physics and chemistry to biology, as well as providing a theoretical basis for the discussion of emergent properties of complex systems. In the words of Glansdorff and Prigogine: ". . . there is only one type of physical law, but different thermodynamic situations: near and far from equilibrium. Broadly speaking *destruction of structures* is the situation which occurs in the neighbourhood of thermodynamic equilibrium. On the contrary, *creation of structures* may occur . . . beyond the stability limit. . . ."

QUANTUM EFFECTS IN BIOLOGICAL PROCESSES

In so far as a biological cell is a factory where many chemical reactions and diffusion flows take place, its behaviour is expected to be analysable in these physico-chemical terms, perhaps even to the extent of explaining the formation of membranes and other cell structures. Nevertheless, biological systems are, at heart, *atomistic and quantised.* Their energy comes in quanta – as in the action of a photon on a chlorophyll molecule, or that of an X-ray quantum in causing a mutation in a DNA molecule – or equivalently, it arrives embodied in certain molecules (ATP) in well-defined excited states. Similarly, the elementary biological processes are operated by individual molecules whose effectiveness depends on a precise composition and structure; even the exact way the molecular chain is folded up is critical. One of the great emergent properties of living systems – the ability of thermodynamically unstable molecules to resist decomposition – may thus be a consequence of quantization, as Schrödinger supposed.[7]

SELF-CONSTRUCTING SYSTEMS

Another great property of living systems is that they are *self-constructing* (as when a chick is formed from an egg, by internal actions directed by self-contained instructions) and *self-reproducing.*[8] It has often been thought that no automaton could have these properties, i.e. that a machine could construct only things simpler than itself. This, however, is known to be false, at least in principle. Turing[9] proved in 1937 that a "universal digital computer", composed of a finite number of parts, was possible which, by scanning and acting on information fed to it bit by bit, from an arbitrarily long tape, was unlimited in its ability to process mathematically expressible information. Later, von Neumann[10] applied this theorem to computer-controlled constructional machines. A Turing computer is made of a *finite* number of parts and so needs a finite amount of information to describe its construction. This information, set down on a tape, could thus be processed by another such computer and if this were designed like a numerically controlled machine tool, so as to *act* on its processed information instead of merely recording its output on paper, then one such computer could construct its fellow. It could even make one more complicated than itself! Needless to say, a computer with such properties would be a very complex system indeed (200 pages of von Neumann's book were needed to describe it!). But the key to its properties lies in a simple consideration: since a system of N parts can in principle have of the order of N^2 distinct binary cross-connections, versatility can increase rapidly with complexity. Thus, von Neumann concluded that there is "a minimum number of parts below which complication is degenerative, in the sense that if one automaton makes another, the second is less complex than the first, but above which it is possible for an automaton to construct other automata of equal or higher complexity". In other words, self-constructibility is an emergent property of a complex system.

SUBJECTIVE ASPECTS OF EMERGENT PROPERTIES

We have already seen, in the case of friction and thermodynamical irreversibility, that some emergent properties result from our own ignorance of or disinterest in detailed information about complex systems. They should not be dismissed lightly on this account, for some of their roots reach down into the deepest foundations of physics. Gold[11] has emphasised the subjective component in many scientific concepts, as follows:

> The basic concepts of physics have evolved from primitive subjective notions about the external world — space, time, force, velocity. They were singled out from other subjective notions because 'objective' measurements were possible ... certain 'laws of physics' could then be defined that described these regularities with very high precision. No such demonstrable success attended the use of any other concepts ... which did not allow themselves to be turned into 'objective' ones through the process of measurement. It is obviously a serious question whether the criterion that 'all observers have to agree' really is sufficient to define objectivity. What if all observers bring the same subjective notion into the measurement? ... The concept of the passage of time is one that does not seem quite to fit into this 'objective' world, and that is not really needed in a definition of the laws that characterise the behaviour of matter. Yet, that time passes is so apparently self-evident that it hardly allows of further discussion ... every object, it is thought, experiences the *flow* of time. ... But which way does the clockmaker make the hands go round? There is nothing objective about that: he fixes the gears so that in *his* appreciation of the flow of time the hands progress from the low numbers to the high. ... If the big pattern of world lines, which contained everything there is, had no indication of any flow, why do we insist on taking this concept into physics, when otherwise we try and free the discussion of physics from subjective 'impressions'?

Thus the flow of time — and we may add to it the related notion of "the present" or "now" — appears to be an emergent property of ourselves, a subjective embellishment of the more austere physical concept of time as no more than a coordinate along which events exist tenselessly. Recent investigations have supported the view, against all intuition, that there is no role for the "flow of time" or "now" in the physical world (for a summary, see Davies[12]). Yet in the mental world these have the unshakeable validity of direct experience. Here is one place, then, where science appears to have brought us right up to the edge between the material and mental worlds and it does look as if there might be an unbridgeable gap between them. In the face of such dilemmas as this, some physical scientists and modern philosophers tend to dismiss the subjective aspects of nature as mere illusions, thus turning their professional backs on clearly dominating aspects of their ordinary lives. Why do they not tell us now the winner of next year's Derby? That would be a quick way to dispel our illusions about the flow of time.

The property *par excellence* of this kind is *consciousness* or *self-awareness*. We have no idea of how consciousness might be explained from physics and chemistry; and, not surprisingly, some people have scorned it as a primitive belief in a "ghost in the machine". But there is perhaps one small constructive step we might take, in the direction of this problem. It is to remind ourselves that our *impressions* of the collective properties of complex matter are often not at all like the elementary processes which constitute them. When we touch a hot surface, we detect a molecular motion but our experience of this sensation is nothing like kinematics. The feeling of *hotness* is a different kind of sensation altogether. Again, when we look at a newspaper picture we see, not an array of dots, but the image of a well-known face. In both cases, we subjectively dredge collective properties out of the complex of our sensations, and these collective properties belong to entirely different categories of experience than their physical causes. A complex system may thus appear to us as something transcendentally different from the individual elements which constitute it. Looked at in this way the disjunction between the singleness of the mind and the complexity of the brain, and with it the great problem of the relation of mind to matter, may become a little less formidable.

ACKNOWLEDGEMENTS

I am grateful to Professor W. H. Thorpe and Dr. C. J. Adkins for useful comments on the draft of this article.

REFERENCES

1. K. G. Denbigh, *An Inventive Universe*, Hutchinson, 1975.
2. M. Eigen, *Naturwiss.* **58**, 465 (1971); *Quart. Rev. Biophys.* **4**, 149 (1971).
3. P. Glansdorff and I. Prigogine, *Structure, Stability and Fluctuations*, Wiley-Interscience, 1971.
4. A. M. Zhabotinsky and A. N. Zaikin, *J. Theor. Biol.* **40**, 45 (1973).
5. A. T. Winfree, *Science*, **181**, 937 (1973).
6. A. M. Turing, *Phil. Trans. Roy. Soc. Lond.* B, **237**, 37 (1952).
7. E. Schrödinger, *What is Life?*, Cambridge U.P., 1951.
8. J. Monod, *Chance and Necessity*, Collins, 1972.
9. A. M. Turing, *Proc. London Math. Soc.* **42**, 230 (1937).
10. J. von Neumann, *Cerebral Mechanisms in Behaviour*, Wiley, New York, 1951.
11. T. Gold, in *Modern Developments in Thermodynamics*, edited by B. Gal-Or, p. 63, Wiley, 1974.
12. P. C. W. Davies, *The Physics of Time Asymmetry*, Surrey U.P., 1974.

R. W. Cahn

Professor of Materials Science and Dean of the School of Engineering and Applied Sciences, University of Sussex.

Recent research has been on the recrystallisation of metals, ordering processes in alloys, formation of metastable phases by ultrarapid cooling of molten alloys. Has built up in the last 4 years a substantial research group concerned with "splat-quenching", and is now increasingly concentrating on glassy alloys made by this process.

TRANSFORMATIONS

Circe, the enchantress, turned men into swine: but the beasts, as they snuffled among the acorns, wept for their lost human forms, for they preserved the minds of men. The change of outer form while some inner essence is maintained intact — the process of *transformation* — is a recurrent theme in literature, mathematics and science alike. In science, the notion of transformation is linked with that of *structure*, for structure determines appearance. The bones of the skull fix the features, the sequence of amino acids specifies the gene that codes for eye colour, the arrangement of atoms or molecules in a crystal determines its shape. Note, however, a crucial distinction between the gene and the crystal. It is of the essence of a gene that it is almost always invariant and replicates precisely true to type; when it does undergo a minute mutation, that in turn becomes invariant, and the resultant biological change with it. Not so with a crystal: the array of atoms is labile and may change reversibly from one pattern to another.

Thus a crystal represents something special in nature, for it is at the same time a single substance and two or more substances. When iron is heated, a well-defined temperature ($910°C$) is reached when the stacking of iron atoms all at once changes; in crystallographic language, it changes from body-centred cubic to face-centred cubic. On cooling, the structure changes back, and this change on cooling brings with it changes in properties (Fig. 1). In pure iron these changes are trivial, but dissolve a small amount of carbon in the hot iron and the resultant alloy — a simple form of steel — transforms with a dramatic change of properties. The previously soft and pliable steel becomes very hard and brittle.

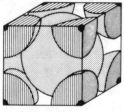
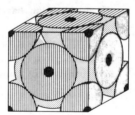

Iron-cold   Iron-hot

The crystal structures of iron below and above $910°C$. Each sphere represents an identical atom.

Not all crystals behave in this fashion. Some transform, sharply, at one or more well-defined temperatures. Some (including steel) transform over a range of temperatures. Some transform suddenly and others take their time. Some crystals can transform only if the pressure to which they are subjected changes: the most celebrated instance of this is the conversion of almost pure carbon in the form of graphite to the form of diamond when it is heated under a very large confining pressure. Still others cannot transform at all; melting, that is, the loss of all crystalline order, preempts any change of crystalline pattern. The generic name for transformations in crystals is "polymorphic transitions"; the Greek word implies the existence of many *forms*. The concept of polymorphism, however, implies that some feature is common to all the forms: there is a unifying essence, just as there is between the grub, the pupa and the moth. What is it?

It is easy to say that the chemical identity of a polymorphic crystal is invariant: but what does the term mean? A chemical substance is defined by the nature and proportions of its constituent atoms and the way they are assembled. Two organic chemicals can be made up of the same atoms in the same proportions in each molecule, yet be put together differently and have quite disparate properties. Such *isomers* are not in general mutually convertible. So the whole molecule is not likely to be the invariant feature in a polymorphic crystal and indeed most such crystals do not contain recognisable molecules at all. The archetypal polymorphic crystal is either an element – such as iron, cobalt, phosphorus, sulphur, uranium – or a simple inorganic compound or solid solution, such as $CaCO_3$, ZnS, $CuZn$, $Cu(Al)_x$. Any such substance forms a crystal that is in effect a single giant molecule: one cannot pick out a single zinc and a single sulphur atom and call that pair a molecule. The only constant feature in a polymorphic crystal is the heap of atoms, the elementary chemical building blocks. What varies when such a crystal is heated or compressed is the nature of the chemical binding; the strength, length and mutual inclination of the chemical bonds changes and the constituent atoms may cluster and rearrange themselves so that the local composition varies from point to point.

The understanding and control of polymorphic transitions is the central concern of the science of metallurgy. This is inevitable, for metallic artefacts have to be formed – and so must be soft and pliable – and they also have to be strong and hard to withstand the shocks and stresses of service. That paradox can only be resolved by transforming the structure of the artefact *after* it has been shaped and put together.

The technological importance of transformations in metals and alloys, then, is evident enough and the scientific problems are subtle and varied in the extreme. To appreciate why this is so, it is necessary to invoke another dimension, that of *microstructure*. Most metallic objects consist not of a single crystal but of an assembly of small irregularly shaped crystal grains, which can only be seen with the aid of a microscope. Generally more than one species of crystal is present and useful alloys are most often composites of several distinct crystalline *phases*. The sizes, shapes, proportions, compositions and mutual disposition of these phases – collectively, the microstructure – are all variable and subject to control. Heat-treatment, designed to alter the microstructure, is the metallurgist's central skill.

The behaviour of a simple carbon steel – the most important industrial alloy – will serve to exemplify the range of phenomena which are covered by that deceptively simple term, *transformation*. When the high-temperature form of steel (right-hand sketch, Fig. 1) is slowly cooled, it breaks up on transformation into two crystalline forms: almost pure iron in the form shown on the left, together with a compound of iron and carbon (Fe_3C) with a more elaborate crystal structure. (The high-temperature form of iron can readily dissolve carbon; the low-temperature form cannot.) Small crystallites of Fe_3C are independently nucleated in many sites. Part of the microstructure consists of thin plates of iron and of Fe_3C in alternation, and the whole assembly is fairly soft. If the high-temperature form is instead cooled suddenly, the transformation process is entirely different. The dissolved carbon is pinned in its existing sites and cannot segregate, for lack of time, and the alloy transforms by an ordered shift of millions of atoms into a new pattern. In the slow process, atom movements are at first random and uncoordinated; in the fast process, they are disciplined and simultaneous. The terms "civilian" and "military" have been applied to the two categories of transformation.

The product of the military transformation, containing as it does a great deal of carbon in enforced solution, is extremely distorted and therefore hard but also incapable of resisting intense shocks. If now this product is tempered by slow progressive heating, a new civilian transformation begins and a succession of iron-carbon compounds is formed in sequence, in the form of minute crystallites. Any desired compromise between hardness and shock resistance can be achieved, and different parts of the same object can be made to have quite different microstructures. The classical Japanese sword represents the most sophisticated application of these skills.

Military and civilian transformations each exist in rich variety, with many subtle distinctions of mechanism. Steels in particular form a large metallurgical family because of the variety of alloying elements which can be added to the basic iron/carbon constituents. Some of these elaborate steels are sensitive to imposed distortion. In such steels, transformations are induced when a hot sheet or rod is forced into a changed shape, as happens when it is passed through a pair of rollers. An early form of this was the use of a special steel — Hadfield's manganese steel — for the construction of railway points; every time a wheel crashed against a crossover point, the point became a little harder because of the stress-induced transformation. The study of such *thermomechanical treatments* is a new chapter in metallurgical research and is at a scientifically most intriguing stage. The strangest variant is the *shape memory effect*. Certain alloys — the alloy NiTi is the best known — can be extensively deformed and then, on heating, will return to their pristine shapes. This behaviour is quite different from that of an elastic spring: a spring obstinately returns to its original shape when it is let go, whereas an SME alloy humbly accepts its pummelling and stays put in its new shape. Only when it is heated does it home to its original form, even against a strong mechanical force seeking to prevent it. This mode of behaviour is always based on a stress-induced polymorphic transformation of the military type, followed by a reverse transformation when heat is applied. It is as though the natural pugnacity of an army forced into precipitate retreat could only be regenerated by an exposure to sunshine!

The shape-memory effect is extremely intriguing, both for its engineering applications and because of the difficulty of understanding the long-range forces which powerfully drive the transformed alloy back towards its original shape. A most detailed investigation of the microstructural changes is necessary in order to come to grips with this phenomenon, which is part of a much larger complex of questions concerned with the interplay of temperature, stress and transformation.

Quite apart from the technological justification, many metallurgists, physicists and chemists have long found this field of study irresistible for its purely scientific attractions. It is a satisfyingly hydra-headed creature: two questions raise their heads for each question that is resolved. For instance, the field of transformations in liquid crystals has arrived in the past few years, as a new branch of physics (stolen while no one was looking from the chemists who opened it up). Liquid crystals are half-crystalline, half-disordered substances that respond sensitively to heat and electric fields: the transformations in liquid crystals from greater to lesser degrees of order have close family resemblances to ferromagnetic and "atomic-order" transformations, and are proving amenable to interpretation in terms of a form of "catastrophe theory" which was applied to ferromagnetism and atomic order long before it was taken up and generalised by mathematicians.

The attractiveness of transformations as a subject for scientific investigation may have something to do with a universal, prescientific human obsession, attested by much ancient legend and folklore. The Greeks told tales of Proteus, a sea-god. Men would seek and seize hold of him as he sat on the rocks, in order to force him to grant a wish. The evasive god would transmute through all the varieties of living appearance, many repellent or terrifying, in the hope of frightening off his captor so that he might escape beneath the waves. Only the brave man who dared to keep his grip on the god till he had run through his entire gamut of forms and returned to his own godlike shape was assured of his wish. Each creature has its proper form and any departure from it is an affront to the natural order of things. A man's sense of identity is indissolubly linked to his own physical body: those who destroy this link, like Circe, like the jealous fairy who turned the prince into a toad, have always been seen as wicked and destructive. To retain the burden of identity while suffering a mutation in appearance is one of man's enduring terrors.

Underlying this is the philosophical dogma that in nature, building-blocks determine structure and structure defines appearance. Man can use a pile of stones to build a cathedral or a bank, but a pile of iron sulphide "molecules" is not subject to man's will and is bound to form a particular kind of crystal, always, with a particular yellowish colour and cubic external form. Snowflakes grow in different patterns according to the change of temperature and humidity, but they are all variations on a strictly constant theme, set by the hexagonal structure of ice. The structure of a simple crystal is implicit in its building blocks — atoms or molecules — and the predetermined forces which bind them together. Complex biological molecules, proteins, are half-way between the determinate simple crystal and the building whose form is subject to man's free will. The same basinful of amino acids can form a multitude of different proteins according to the template provided. It is still a matter of atomic fit: Monod showed how one amino acid selects another out of a copious mixture and by this means the acids assemble into a preordained pattern. As Monod puts it: "The epigenetic building of a structure is not a *creation*; it is a *revelation*." This is as true of an element as it is of a protein.

Polymorphic transformations (and indeed the melting of crystals) thus conflict with a deep philosophical sense of fitness, of match between form and essence. Those crystals which transform easily are unstable, not sure of their own proper nature, full of doubt, like man tormented by his Freudian unconscious. It is ironic that the most protean of elements is also the most unstable and dangerous: plutonium has four polymorphic forms. By gaining an understanding of transformations, man masters his fears, keeps hold of Proteus, gains his desires, fulfils his needs. At this level, the philosophical, scientific and technological desires of man all impel him to the same study. The mastery of transformations reassures man, the magician, that Jekyll and Hyde are under control.

Sir Denys Haigh Wilkinson, F.R.S.

Vice-Chancellor of Sussex University from September 1976. Formerly Professor of Nuclear Physics and Experimental Physics, University of Oxford.

Hughes Medallist of the Royal Society, London, 1965. Knight Bachelor 1974.

Founder Member of the Governing Board of the National Institute for Research in Nuclear Science 1957-65; Chairman of the Nuclear Physics Board of the Science Research Council 1965-8; Chairman, Physics III Committee and Member of the Scientific Policy Committee, CERN, Geneva, 1971-5.

THE UNKNOWN ATOMIC NUCLEUS *

Since the discovery of the neutron in 1932 we have said that the atomic nucleus contains neutrons and protons. More cautiously we should have said that when neutrons and protons (nucleons) are brought together they coalesce to form atomic nuclei. The two statements are not necessarily the same: the nucleons, on coalescing, might lose their identity and form some sort of undifferentiated nuclear "black hole" matter, without granular structure, characterised just by a few overall quantum numbers such as electrical charge and angular momentum. The fact that when nuclei are struck violently together nucleons come out does not prove that nucleons were inside in the first place — barks come out of dogs but that does not prove that dogs are made of barks.

It is, indeed, very difficult to get anything like direct, as opposed to inferential, evidence that there *are* nucleons inside nuclei. Perhaps the nearest approach to direct evidence comes from bombarding heavy nuclei such as lead with very energetic protons, of about 25,000 MeV in the best (CERN) experiments, measuring the energy distribution of the outgoing protons closely in the forward direction and comparing that distribution with what would be expected if the nucleus were indeed composed of so many protons and so many neutrons that interacted individually with the bombarding proton just as they would in the free state. The theory, put like that, sounds easy but it is extremely complicated to work out in full numerical detail because you have to allow for the possibility that the bombarding proton makes not just one but many successive collisions as it struggles through the nucleus and also for the possibility that it creates secondary particles (mesons) as it goes. However, when theory (the so-called Glauber method as worked out by O. Kofoed-Hansen) and experiment are compared they agree very well (better than 30 per cent) over the whole investigated energy range which is about as good as could be expected in view of the inaccuracies in our knowledge of the input data themselves, viz. the free-space nucleon-nucleon collision probabilities.

Convinced that nuclei contain nucleons pretty much as in the free state our next question is: what are those nucleons *doing*? The answer is a surprisingly simple one, namely that the nucleons are just cruising around essentially independently of one another as if any given nucleon were moving in some central field of force that must obviously, in the last analysis, be due to the fellow nucleons of the one in question but into whose origins and detailed nature we do not need to inquire for an initial discussion of the motion of the given neutron or proton. Of course, the motions of the several nucleons must be correlated to some degree because neutrons and protons are fermions and so obey the Pauli exclusion principle — each nucleon must occupy a separate quantum state: just as with electrons in an atom or in a metal every quantum state from the energetically lowest is occupied until all the nucleons of the nucleus are accommodated; the ensemble forms the nuclear Fermi sea but the associated correlations between the nucleons of the sea are due to the laws of quantum mechanics, the Pauli principle, the anti-symmetrisation of the overall wave function, and do not in the slightest imply that any forces explicitly operate between the particles even though we know that such forces must exist because there is nothing else available to bind the nucleons together into the nucleus; cf. the situation in an atom where we similarly build up our zero-order picture of atomic structure by ignoring the forces between the electrons but where the overall field of force within which the electrons move is primarily due to an external agency — the central atomic nucleus.

We can easily determine the depth of the nuclear Fermi sea because it must be equal to the depth of the effective potential within which the nucleons are moving (and that we can infer from the way in which bombarding nucleons are refracted as they enter and leave nuclei) minus the energy with which the topmost nucleon of the Fermi sea is bound into the nucleus (which is just the least energy that we must pay into a nucleus to persuade it to release a nucleon); the answer is about 100 MeV.

We know that this picture of nucleon motion in independent non-interacting orbitals – the zero-order nuclear shell model – is reasonably correct because we can knock, say, protons out of a nucleus by bombardment with, say, energetic electrons (500 MeV or so in the best (Saclay) experiments) – an $(e, e'p)$ reaction – and thereby – through the e vs. e', p *energy* imbalance – map out the energy distribution of the protons within their Fermi sea and also – through the e vs. e', p *momentum* imbalance – map out the associated momentum distribution of the protons which is linked by general laws of quantum mechanics, having nothing in particular to do with the nucleus, to the *angular* momentum distribution of the protons, this latter being itself specified by our independence, shell model, hypothesis. Energy and (angular) momentum distributions tally most impressively with expectation in light nuclei such as ^{12}C.

How can we understand this remarkable simplicity, this atomic-like structure, of the nucleus? The simplicity of the atom occasions no surprise because the electron-electron interaction, that perturbs the simple orbital motions of the individual electrons, is feeble compared with the interaction of each electron with the central nucleus that produces the simple orbital motions. But in the nucleus the nucleon-nucleon forces that tend to *break down* the simple orbital motions (nucleons "bumping into each other") are the very same that must *give rise to* the overall field of force that somehow generates those simple orbital motions. So how can the motions be so simple? The answer to the conundrum lies in the Pauli exclusion principle that gives rise to the notion of the Fermi sea in the first place. Imagine that the nucleon-nucleon interaction is zero and that the nucleons' orbital motions are established by some external field of force so that every quantum state up to some maximum is occupied by a single nucleon. Now switch on the nucleon-nucleon interaction that will try to make nucleons bump into each other and so break down the simple motions of the Fermi sea; the "bump", however, will not be effective unless it is sufficiently violent to lift *both* participant nucleons right out of the Fermi sea into unoccupied quantum states above its surface since smaller excitations would put the bumping nucleons into already-occupied states and this is forbidden – one state, one nucleon. So whether or not the simple Fermi-sea motions survive in a real nucleus is a question of the depth of the Fermi sea (100 MeV or so as we have seen) in relation to the interaction energy between two nucleons at their usual spacing inside the nucleus. It is not too easy to give a simple answer for this interaction energy because it depends on the details of the nucleon motions and on the effective spectroscopic states in which the "bump" takes place but it is certainly only of the order of 10 MeV. It is therefore not surprising that the Fermi sea survives except in the immediate vicinity of its surface where excitations into a few higher states may take place: the "configuration mixing" that distinguishes the modern shell model from its primitive forebear of 1950.

Turning to the quantification of these ideas, we find that detailed shell-model calculations of nuclear structure, starting from nucleons whose orbitals are adjusted in size to fit actual nuclear dimensions, but *including* the experimentally determined nucleon-nucleon interactions operating *between* the nucleons, give quite good agreement with experiment not only in the level schemes but in static properties such as magnetic moments and in dynamical properties such as radiative transition probabilities. Technically much-more-elaborate unrestricted self-consistent calculations, similar in principle to Hartree-Fock calculations of atomic structure, in which the nucleons are left to work out their own salvation on the basis of their mutual interactions, without the presupposition of any particular dimensions for the resulting orbitals, yield theoretical nuclear dimensions, surface thickness and binding energies that agree quite well with nature. An additional feature of these Hartree-Fock calculations is that they predict that many nuclei should be not roughly spherical but rather strongly deformed — half as long again as they are wide in some cases — and this is in striking accord with experiment.

Many nuclear properties can be simply represented by speaking in classical-sounding terms of the bulk properties of the nucleus: it rotates as a whole or vibrates as a whole. Such collective descriptions of nuclear behaviour are completely consistent with the microscopic shell model that we have just sketched and may be directly derived from it — in full quantitative detail in sufficiently simple cases and in convincing outline in more complicated ones where we believe we are held back from a full description only by lack of adequate computer power.

All this we shall call the "conventional" approach and it makes it sound as though nuclear structure is a closed book — we measure empirical forces between nucleons, we calculate, on the basis of those forces, the properties of complex nuclei and the answer comes out right. This is true, but only up to a point.

There are two problems. The first is that our detailed accounts of nuclear structure really involve only a few "valence" nucleons near the top of the Fermi sea — just a few per cent of the total nuclear contents. True it is that these few nucleons turn out experimentally to have the quantum numbers we should expect of them if all those below them in the Fermi sea were doing what the shell-model would have them do: but we have no evidence that the lower nucleons are in fact behaving in this "conventional" manner. It is unlikely that the quantum numbers of the valence nucleons would come out right if the under-pinning nucleons were behaving radically differently from the "conventional" prescription, e.g. were set into a crystalline lattice. On the other hand, it is quite possible that, to continue to use "conventional" language, the appropriate admixture of configurations lying *above* the valence nucleons into those lying *below* could transform the simple shell-model orbital motions of some at least of the under-pinning nucleons into quite tight self-centred clusters such as alpha-particles. In this case a gentle probing of the nucleus, involving only rearrangement of the valence nucleons among themselves, would give the "conventional" answer but a more vigorous probing may reveal a predilection of the nucleus to fall apart into what we should then probably have to take to be literally pre-formed clusters: we should begin to think of the nucleus as a raspberry; there are some signs that this may be so; we must keep an open mind. The second problem is that when the "conventional" approach is pressed really quantitatively in a place where it should be able to withstand that pressure, i.e. in a place where, in its own terms, its predictions are unambiguous and not qualified by uncertainties in its own formulation or its own relevant parameters, it fails.

A single immediate illustration of this quantitative failure of the "conventional" approach will suffice. The simplest nuclear reaction is the radiative capture of slow neutrons by protons to form deuterons: $n + p \rightarrow d + \gamma$. The known nucleon-nucleon forces permit an accurate and unequivocal prediction of the cross-section for this process: $(0.303 \pm 0.004) \times 10^{-24} \, \text{cm}^2$. The experimental number is also accurately known: $(0.332 \pm 0.002) \times 10^{-24} \, \text{cm}^2$. The discrepancy between theory and experiment is therefore (10 ± 1.5) per cent. We find similar discrepancies between theory and experiment wherever we look in systems of the type where the "conventional" approach has no escape: the β-decay of ^{3}H and other light nuclei; magnetic moments of nuclei of supposedly well-known structure; the density of nuclear matter; the binding energy of nuclear matter. Something almost always goes wrong at the few per cent level: what is it and can we put it right? To see what is wrong in the "conventional" approach we go back to the beginning: we have spoken of empirically determined forces acting between nucleons as the basis of our theories of complex nuclear structure; but where do those forces come from and should we not have regard to their origins in working out the consequences of their actions? In other words the properties of nuclei may depend not only on the forces between the constituent nucleons but also on the means by which those forces are engendered.

Before we look at the forces between nucleons let us look at the individual nucleons. Nucleons may or may not have an explicit sub-structure of quarks, and that is not for our present consideration. But, in any case, the nucleon must be thought of as a source of virtual mesons: just as the electron is surrounded by its electric field, the quantization of which is virtual photons, a tenuous cloud of which must therefore be thought of as surrounding any electric charge, so a nucleon is surrounded by its nuclear field, the quantization of which is virtual mesons, a cloud of which surrounds the nucleon. But since the meson-nucleon coupling is much stronger, by a factor of 100 or so, than the photon-charge coupling, the meson cloud around a nucleon must be pictured as dense rather than tenuous, with several virtual mesons around at any moment, so that the meson cloud dressing a nucleon is to be thought of as part of the normal essential substance of the nucleon and not, as in the case of the virtual photon dressing of an electric charge, an insubstantial frill that may be disregarded to a first approximation.

Indeed, when the internal charge distribution of a proton is probed by the scattering of energetic electrons from it, it is found that the proton, unlike the electron, is not a point charge but an extended structure of $\langle r^2 \rangle^{1/2} \simeq 8 \times 10^{-14}$ cm. Now mesons are particles of finite mass m so that to bring them into virtual being costs energy of at least $\Delta E = mc^2$ and therefore the time Δt for which they can enjoy their virtual existence is limited by Heisenberg's uncertainty principle to $\Delta t \cdot \Delta E \simeq \hbar$, viz. $\Delta t \simeq \hbar/mc^2$: they will therefore not be able to stray to distances greater than about $c \cdot \Delta t \simeq \hbar/mc = 1/\mu$ which distance sets a natural spatial scale for the nucleon through a factor of form $e^{-\mu r}$. Now for the lightest meson, the pion, $mc^2 \approx 140$ MeV, $1/\mu \approx 1.4 \times 10^{-13}$ cm; heavier mesons, with correspondingly smaller values of $1/\mu$, are legion (for example, the strongly coupled ρ- and ω-mesons have $1/\mu \simeq 3 \times 10^{-14}$ cm) and it is not surprising that the whole mesonic dressing should add up to the observed $\langle r^2 \rangle^{1/2} \simeq 8 \times 10^{-14}$ cm.

So nucleons may be thought of as fuzzy balls of size about 10^{-13} cm; when two nucleons approach, each has no means, apart from the relatively weak electric force that we disregard at this stage, of knowing that the other is there until their mesonic clouds begin to interpenetrate; when this happens we can imagine that occasionally a meson that has emerged from one nucleon will not go back in again but will rather pass across to the second nucleon and thereby carry word of the first one's presence, i.e. establish an interaction, a force. Again, we should expect this force to display the factor $e^{-\mu r}$.

It is therefore this exchange of mesons, pions at the greater distances, heavier mesons, pairs of pions and so on at smaller distances, that constitutes the force between nucleons that we can empirically measure and that, then forgetting the mesonic origins of the force, we use as the starting point of our "conventional" theories of nuclear structure — that typically get things wrong by about 10 per cent. However, the overall properties of a complex nucleus must depend on the totality of what is going on inside it and not just on what its nucleons are doing; for example, the flow of charged pions between the nucleons, establishing the force between them, constitutes an electric current in addition to the currents due to the motions of the protons themselves and so will contribute to all current-linked phenomena, for example the magnetic moment. Another possibility is that a pion "in the air" between two nucleons may disappear to give an electron-neutrino pair, $\pi \rightarrow e + \nu$, thereby contributing to the weak (β-decay) properties of the nucleus. Such mesonic effects must show up everywhere to some degree and when we calculate that degree to the best of our present ability, it turns out to be typically a few per cent. We must obviously "mesonate" our "conventional" nucleons-only approach and let in the mesons explicitly.

"Mesonation" has further consequences: when real pions bombard nucleons in the laboratory we find that the nucleons can be thereby raised into a multitude of excited states, generically called resonances or isobars, just as the quanta of the electromagnetic field, photons, can raise systems held together by that force, e.g. atoms or molecules, into excited states. So when nucleons interact, which they do by bombarding each other with pions and other mesons, in the virtual state to be sure but that makes little essential difference, there must be a certain chance that that bombardment will result in the excitation of isobars. This has two important consequences for our discussion of complex nuclei:

1. If an isobar is formed in a free nucleon-nucleon (N-N) low-energy collision it cannot persist because it is more massive than the nucleon and so, by the Heisenberg uncertainty relation, must disembarrass itself of its excitation energy ΔE within a time Δt given by $\Delta E \cdot \Delta t \simeq \hbar$; the lightest isobar is the Δ for which $\Delta E \simeq 290$ MeV so the longest Δt is about 2×10^{-24} sec; this de-excitation must be effected by, for example, the re-emission of a pion such as gave rise to the excitation and its reabsorption by the other nucleon; this merely constitutes part of the N-N force such as we use in the "conventional" shell model computation. If, however, the two nucleons in question are part of a complex nucleus another option is open: the pion that de-excites the isobar may be absorbed by a third nucleon thereby constituting a force of an essentially 3-body nature that comes into play only when three nucleons are close enough together and that is distinct from the sum of the various possible 2-body forces that operate between the three. This 3-body N-N-N force must therefore be introduced into our computation of nuclear structure additionally to the 2-body N-N forces of the "conventional" approach. (Such genuine N-N-N forces can arise in other ways: for example, the exchange of a ρ-meson between two nucleons is part of the N-N force; but the ρ-meson, after emission by one nucleon of a complex nucleus, may decay, $\rho \rightarrow \pi + \pi$, and not be absorbed as a ρ-meson by a second nucleon at all; if both decay pions are then in fact absorbed by the second nucleon this is again just a piece of the N-N force but if one decay pion is absorbed by the second nucleon and the other by a third we have an N-N-N force.)

2. The isobar, having been formed, may de-excite to a nucleon not by emission of a pion but by emission of, for example, a photon, real or virtual, or an electron-neutrino pair, which will therefore constitute an electromagnetic or weak interaction of the nuclear system additional to that involving nucleons alone (or nucleons plus mesons only).

A further possibility is that a pion, in flight between two nucleons A and B, may dissociate into a nucleon and an anti-nucleon; the anti-nucleon then annihilates with nucleon B giving a pion that is absorbed by nucleon A so that the final state, as the initial one, consists just of two nucleons and this process is just part of the N-N force. However, if nucleons A and B are part of a complex nucleus, the pion coming from the annihilation of the anti-nucleon with nucleon B could be absorbed not by nucleon A but by a third nucleon of the initial system thereby constituting another component of the N-N-N force. Similarly, the annihilation could yield a photon or an electron-neutrino pair rather than a pion so that this pion-dissociation mechanism contributes also to the electromagnetic and weak properties of the nucleus.

What is the quantitative importance of these processes whose qualitative role is undeniable? We examine the example of radiative neutron-proton capture that we saw above to have a (10 ± 1.5) per cent discrepancy with "conventional" nucleons-only theory. We now see that the "conventional" theory in which the gamma-ray "comes out of the nucleons" must be supplemented by at least three additional mechanisms:

(i) a pion is emitted by one nucleon but, while it is "in the air", before being absorbed by the other nucleon, emits the gamma-ray;

(ii) a pion is emitted by one nucleon and on absorption by the other raises it into an isobaric state that de-excites emitting the gamma-ray;

(iii) a pion is emitted by one nucleon and dissociates "in the air" into a nucleon and an anti-nucleon; the anti-nucleon annihilates with the second initial nucleon giving the gamma-ray.

The sum of these three mechanisms in fact completely removes the 10 per cent discrepancy. Other discrepancies among the properties of the lightest nuclei, for which we think we have quantitatively-reliable "conventional" predictions, are similarly removed, or significantly reduced, by similar mesonic interventions.

It is therefore quite wrong just to think of the mesons as the generators of the N-N force then to use that force for nuclear structure computations and forget about the mesons that gave rise to it: they must be brought in *explicitly* if we want to get the right answer.

So far we have spoken chiefly of pion-exchange because the pion is the lightest meson and so has the longest "reach": $1/\mu = \hbar/mc \simeq 1.4$ fm. But the heavier mesons can also be exchanged and give important contributions to the N-N force at smaller distances; for example, the ρ-meson's coupling to nucleons is very strong and although its reach is only about 3×10^{-14} cm it is an important contributor to the N-N force. However, the fact that a meson is heavy does not exclude it from certain long-range effects. For example, a ρ-meson could emerge from a nucleon, quickly convert to a π-meson by emission of a photon or interaction with the electromagnetic field, $\rho \rightarrow \pi + \gamma$, giving a "long-reach" pion to be absorbed by another nucleon. This process contributes to certain nuclear properties, specifically to the sum of the magnetic moments of nuclei having complementary numbers of neutrons and protons, e.g. ^{3}He and ^{3}H, to which the process $\pi \rightarrow \pi + \gamma$ does not, and so it can be, and has been, picked out and identified.

Complex nuclei must similarly contain "strange" mesons and hyperons by virtue of processes such as $\pi \rightarrow \Sigma + \bar{\Sigma} \rightarrow \pi$ and $N \rightarrow \Sigma + K \rightarrow N$ but it is more difficult to put their undoubted presence into evidence.

Although it is now clear that explicitly mesonic effects are important at least at the 10 per cent level this does not mean that the procedures by which we have introduced them are adequate or even right. We have taken the line that we start from an empirical N-N force that is presumably generated by meson exchange in a manner that we do not understand in detail; the mesons are then forgotten for the time being, the empirical N-N force being used to calculate the structure of the "conventional" system of nucleons only, moving under the influence of that N-N force; such mesons as we know how to treat adequately are then let back into that nucleonic structure without changing it and their explicit additional effects such as their electromagnetic and weak properties and their generation of isobars and N-N-N forces are essentially treated as perturbations that do not fundamentally change either the initial structure or the nucleons that constitute it. We will now look at a few difficulties.

The first difficulty is that this "mesonated conventional" approach is essentially ambiguous, for example we quickly find that we are running the danger of double counting: when long-lived nucleon states, such as the zero-order shell model orbitals of the Fermi sea, are involved it is fairly clear that the mesonic exchanges between them, of time-order $\hbar/mc^2 \simeq 5 \times 10^{-24}$ sec for a pion and shorter for other mesons, are relatively fleeting additions that can be explicitly separated out and discussed; when, however, we are considering more complicated nucleon motions such as are associated with the successful N-N "bumps" that *do* occasionally lift nucleons out of the otherwise tranquil Fermi sea and that may involve excitations of $\Delta E \simeq$ 100 MeV or much more, we see that such nucleonic excitations are themselves fleeting ($\Delta t \simeq \hbar/\Delta E \simeq 7 \times 10^{-24}$ sec when $\Delta E = 100$ MeV) and are comparable in time scale with the exchange times of the mesons themselves that gave rise to the "bump". The meson exchange and the associated nucleon motion can no longer be cleanly separated from each other and we clearly must not simply add their separately computed effects because they must to some degree just be different representations of each other.

In principle the way out of this first difficulty is to abandon the "mesonated conventional" approach, in which we take the empirical N-N interaction as the starting point and add explicit mesons and isobars later, in favour of a true many-body relativistic field theory in which nucleons, mesons and isobars partake with equal right in an overall description of the resultant system whose properties, such as its interaction with the electromagnetic field, would then be given unambiguously; there would be no forces as such in the input data but only the identity of the various particles and the associated coupling constants appropriate to the various particle combinations.

This counsel of perfection cannot yet be followed but we might hope for progress along the following lines: take a very simple model, soluble in relativistic many-body field theory, a system containing perhaps one species each of nucleon, isobar and meson then: (i) compute the N-N force that would arise from the meson exchanges, use this force in the "conventional" way to calculate "nucleon-only" wave functions, "mesonate" this "conventional" structure by adding the mesons and isobars perturbatively as sketched above and then compute whatever properties you are interested in using various recipes for resolving such ambiguities as you can spot; (ii) compute the same properties from the proper full solution.

In this way we should get a feel for the right way to handle the ambiguities and for the degree to which the "mesonated conventional" approach can be relied upon as a stand-in for the real thing. Although it may not be possible to treat field theoretically a realistic assembly of mesons, nucleons and isobars it may be possible to treat separately several sub-sets, viz. different limited selections of mesons and isobars, and so get a feel for whether the ambiguities can be systematically resolved or whether different recipes are going to be necessary for different meson exchanges, etc.

The second difficulty of the "mesonated conventional" approach is related to the first and is that it may miss entire new modes of nuclear behaviour essentially because it closes its eyes to the possibilities. For example, our "mesonation" considers only meson exchanges between pairs of nucleons, or threes when isobar excitation is included. But it is entirely possible that mesonic circuits involving more than three nucleons may get excited with the very strong $\pi + N \to \Delta(N\Delta\pi)$ interaction, involving the Δ-isobar, as a kind of catalyst for the generation of pions. Indeed, so strong is the $N\Delta\pi$ interaction that it is quite on the cards that the *entire nucleus* may participate in such pionic circuits with the pions then pursuing orbits referred to the nucleus as a whole like those of the nucleons themselves. The total energy of the system could then minimize at a very high pion density: the so-called "pion condensation" phenomenon, associated particularly with A. B. Migdal, that has been much discussed recently. Although it seems unlikely that nuclei of normal density already contain such condensed pions (a point made especially by G. E. Brown) it also seems likely that this phenomenon might set in at densities elevated by only a factor of two or three above normal, so that real nuclei may already contain the beginnings of such condensations, on a short time scale, associated with density fluctuations; these mini-condensations or multi-nucleon pion circuits may already have some importance for the overall energy balance and, specifically, might lower the energy of certain types of nucleon cluster states, viz. promote nucleon clustering into more-or-less well-defined nuclear sub-structures the possibility of which we noted earlier; these sub-structures would not necessarily be long-lived but might be numerous and even fairly well defined if seen in a snapshot.

If quasi-stationary sub-nuclear many-$N\Delta\pi$ clusters exist then the Δs themselves in such clusters will be modified from their free-space properties just as the properties of a coupled system of identical resonators of any kind differ from the properties of the individual identical resonators: the effective energy and width of the Δ will depend on the type of cluster of which it is part. This we might hope to see by exciting such clusters through the bombardment of nuclei by real pions: in principle the Δ-resonance might shift or split and we should probably stand our best chance of seeing this by looking specifically in final-state channels that might be preferentially fed in the ultimate decay of such clusters, viz. involving high-energy complex fragments such as α-particles, etc., as has been emphasised by E. Vogt.

Even more radical "whole-nucleus" consequences of mesonic exchanges might be envisaged. Although the $N\Delta\pi$ interaction that may make pion condensation possible is very strong it is a p-wave interaction that peaks for free pions bombarding stationary free nucleons at $E_\pi \simeq$ 200 MeV so that the individual condensed pions tend to be "energetic" and this tends to raise the energy of the condensate. However, consider the case of the hypothetical $J^\pi = 0^+$ σ-meson that would have a strong s-wave interaction with nucleons; this could promote condensation for "zero-energy" σ-mesons and it has been suggested by T. D. Lee and G. C. Wick, on the basis of a σ-N many-body field-theory, that the associated nucleon density could be very high — perhaps 10 times normal — and the system more stable than the normal nuclear ground state. It is therefore conceivable that there are some such collapsed nuclei, "abnormal nuclear states", already around in nature but if so they are very few. Recent work by M. Rho suggests that although such abnormal states of high density can probably be made they are probably also high in excitation above the normal ground state and so will be difficult to identify. However, their transitory role, riding upon density fluctuations in ordinary nuclei as discussed above for pions, may not necessarily be totally negligible. Note that even if an identifiable σ-meson does not exist as such (and a broad structure of the correct quantum numbers has been seen in the $\pi\pi$ system at a total energy of about 600 MeV) states of appropriate properties could be synthesised out of pairs of pions, $K\bar{K}$, etc.

An interesting question of practical importance is raised by these field-theoretical and other discussions of the density of mesonic fields as a function of density of nucleons: the time scale. If the nucleon density is suddenly changed, say, in a shock-wave generated by the collision of two fast heavy ions or in a spontaneous nuclear fluctuation, *how long* does it take, say, the pion field to establish its equilibrium value, *how long* does the phase change to the condensed state, if there is to be one, take to come about? The calculations of, say, the pion field density are equilibrium calculations in which, as a function of nucleon density, pions have themselves a right to a certain equilibrium density by virtue of the magnitudes of the various coupling constants; but these equilibrium calculations do not tell us how in detail the pions *get there*: obviously, in a sense they must "come out of the nucleons" but the calculations do not tell us how and so they do not tell us how long equilibrium will take to reach. The situation is analogous to that of the Planck spectrum of black-body radiation: if I suddenly heat up a cavity how long does it take for the Planck spectrum to establish itself? The spectrum is an equilibrium spectrum to which the cavity has a right determined only by its temperature and not dependent on the material out of which its walls are made and yet the photons must "come out of the walls" and take a time to do it that will be determined by the nature of the walls.

This question of time scale is obviously of importance in the nuclear case because the only means we have available, at least on Earth, for changing the nuclear density are themselves of short time scale; it is quite possible that critical densities for condensation might be exceeded in a heavy ion collision but the nucleus pulls itself apart before condensation has time to establish and manifest itself: that would be a pity.

Another radical possibility is that the many-body forces that are effectively induced by the more generalised sorts of meson exchange that we have just been considering may be such that large nuclei actually become *crystalline*, at least in their interior. This would be an extreme form of cluster structure and there is nothing specific that suggests that it may be happening but so long as we remain ignorant about the organisation of the depths of large nuclei we must keep an open mind. (It may well be that this possibility could be eliminated, theoretically, by arguments based on the shell-model quantum numbers of the valence nucleons — see our earlier remarks about cluster models — or, experimentally, by the Glauber analysis of high-energy p-nucleus scattering which, as we saw, gave good agreement with experiment for an essentially uncorrelated neutron-proton nuclear gas; but neither of these exercises has been carried through.)

Returning to the difficulties of the "mesonated conventional" approach we find a third substantial problem that takes us right back to our discussion of *nucleon* structure. The free nucleon is a dynamical structure containing pion and other mesonic currents. Nucleons interact with each other only by virtue of the exchange of mesons such as contribute to the build up of their individual structures; in saying this we must additionally bear in mind that mesons are not conserved so that the exchange of mesons *between* nucleons will affect the flow of mesons *within* the individual nucleons but not in any simple way. The situation is very much like that of the near-zone of a radio antenna; energy flows out of the antenna and back in again (circulation of virtual photons); when a second, identical, antenna is brought near the first energy now flows out of one and into the other but the original flow of energy out of each and back into the same antenna again is also changed. Thus those properties of the individual nucleons that are affected by their internal mesonic constitution (the *intra*-nucleonic meson currents) such as their magnetic moments and intrinsic β-decay rate will be changed when they are brought into each other's vicinity, in a manner additional to the contribution to the overall nuclear property in question that is due to the *inter*-nucleonic meson currents. Indeed just as in the analogy of the antennae one cannot separate the "intra" from the "inter" effects in any unambiguous way; we find ourselves, at this deeper level, facing the same kind of problem as arose in our discussion of the exchange-current/configuration-mixing ambiguity; in the present case we clearly need a model of *nucleon* structure before the resolution can be effected and this is one stage beyond the relativistic many-body field theory that constituted our in-principle hope for resolving the earlier problem. It is impossible to guess how important such "nucleon modification" effects might be but they will clearly be greater for processes that require nucleons to be in strong interaction and we might hope to get some indication from, for example, high-energy photodisintegration that will depend on, among other things, the magnetic dipole moments of the "individual nucleons".

We have seen how essential it is to take explicitly into account the mesonic structure of the nucleus if we want to get "ordinary" questions answered correctly to better than 10 per cent or so and how far we are from an unambiguous clearing-up of that 10 per cent, a satisfactory treatment of which will have to await at least a semi-realistic many-body field theory. We have also seen how mesonic effects of a type as yet only dimly envisioned may have far-reaching consequences for nuclear structure by, for example, promoting clustering through incipient condensation phenomena. The two messages are distinct but equally clear: (i) even conventional nuclear structure physics already demands "mesonation" for an acceptable understanding; (ii) the new types of phenomena that mesons may engender are so radically different from the expectation of conventional nuclear structure physics that we must keep a very open eye and mind indeed.

Abdus Salam, F.R.S.

Professor of Theoretical Physics at Imperial College, University of London and Director of the International Centre for Theoretical Physics, Trieste.

Awarded the Hopkins Prize (Cambridge University) for the most outstanding contribution to Physics during 1957-8, the Adams Prize (Cambridge University), 1958, first recipient of the Maxwell Medal and Award of the Physical Society, London, 1961 and recipient of the Hughes Medal of the Royal Society, London, 1964. Awarded the Guthrie Prize and Medal, 1976.

Member of the Institute of Advanced Study, Princeton, 1951. Elected Fellow of the Royal Society, 1959; Fellow of the Royal Swedish Academy of Sciences, 1970; Foreign Member of the USSR Academy of Sciences, 1971 and Honorary Fellow of St. John's College, Cambridge, 1971. Chairman of the United Nations Advisory Committee on Science and Technology, 1971-2.

Published widely on aspects of the physics of elementary particles and the scientific and educational policy for Pakistan and developing countries.

PROBING THE HEART OF MATTER

What is Matter? What are the laws which govern its behaviour? Do there exist fundamental entities of which all matter and all energy is composed? These are the quests of Particle Physics — the frontier discipline of Physical Science, an area of intense search, profound ignorance, and also of some of the deepest of man's insights into the workings of nature.

The historical approach to the subject centred on the search for the most elementary constituents of matter — a search whose tenor was so remarkably predicted by Isaac Newton, "Now the smallest of particles of matter may cohere by the strongest attractions, and compose bigger particles of weaker virtue; and many of these may cohere and compose bigger particles whose virtue is still weaker and so on for diverse successions, until the pregression ends in the biggest particles on which the operations in Chemistry and the colours of natural bodies depend, and which by cohering compose bodies of a sensible magnitude.

"There are therefore agents in nature able to make the particles of bodies stick together by very strong interactions. And it is the business of experimental philosophy to find them out."

In Newton's day, no "small particles" were known, and the only "virtue" bulk matter was definitely known to possess was the gravitational force of attraction, a long-range force between two objects of masses m_1 and m_2, at a distance r from each other, which acted according to the law

$$F = \frac{G_N \, m_1 \, m_2}{r^2} \text{ (with } G_N \text{ the Newtonian constant).}$$

The first of "small particles" to be discovered was the electron in late nineteenth century, followed by the proton, the neutron and the neutrino — all discovered by the year 1934. In 1934, these four appeared to be the ultimate constituents of which all matter might be composed. They appeared to possess three further "virtues" besides the gravitational. To characterise these "virtues" quantitatively one ascribes "charges" to these particles; the electron and the proton carry fixed (equal and opposite quantities (e and $-e$) of) electric charge; the proton and the neutron carry equal "nuclear charges", while all four (the electron, the neutrino, the proton and the neutron) carry a "weak" charge in addition to the "gravitational charge" (this latter is better known as "mass"). The "nuclear" and "weak" forces are short range (of the form

$$g_1 g_2 \frac{e^{-r/r_0}}{r^2}$$

where g_1 and g_2 are the magnitudes of charges carried by the relevant particles) in contrast to the electric and the gravitational forces, which are both long range (of the form $e_1 e_2/r^2$ and $m_1 m_2/r^2$). The effect of nuclear and weak forces is felt predominantly when the participating particles are closer than $r_0 = 10^{-13}$ and $r_0 = 10^{-16}$ cm respectively.

To complete the story of these charges, in 1934, Dirac added to the subject the concept of "anti-particles"; every particle has as a counterpart an anti-particle which has the same mass (gravitational charge) but an electric, weak or nuclear charge which is equal in magnitude but opposite in sign to that of the particle.

The subsequent history of particle physics is the history of proliferation and discovery of other objects equally as small in size — or in some cases smaller still — than the four particles mentioned above, so that the suspicion has arisen that possibly these four — or at least some of these four — may themselves be composite of some more basic constituents — "quarks" — not so far experimentally isolated as free particles. In fact the whole concept of "elementarity" — and the notion of "basic" constituents — may have become untenable. A more profitable way of summing up the new discoveries appears to be the statement that as one has probed deeper into the structure of matter (and the standard method for this is collisions of higher and higher energy beams of protons, neutrons, electrons and neutrinos with each other and the study of their reaction products) — one has discovered new "charges", besides the ones mentioned above.

At present some nine charges may be distinguished; there are three so-called "colours", two "isotopic" charges, "strangeness", "charm", "baryon" and finally "lepton" charge. The weak, the electric and the nuclear charges are made up from these nine basic charges. And further — and this is where the biggest excitement of 1973-4 has centred around — we have discovered that there exists a basic symmetry principle, which states as a unifying principle *that all these charges are on par with each other*; that (barring the gravitational charge) all other charges are aspects, are manifestations of one basic charge. The result seems experimentally borne out for weak and electromagnetic charges. The sorts of questions which arise are the following. Do there exist nine basic objects each carrying one single "charge"? Is there anything sacred about the number *nine* for the "basic" charges? Or indeed, is there really a finite number of charges — or shall we go on discovering newer and newer charges as we probe deeper and deeper into the heart of the particles we create in high-energy collisions?

To summarise then, the basic dilemma of our subject is the "charge" concept and its nature. There is one charge — the gravitational (mass) — which we believe is rooted within the concepts of space-time. Einstein identified gravitational force with the curvature of a continuous space-time manifold. Are the other charges, weak, electromagnetic, and nuclear, equally rooted within space-time, perhaps giving an indication of its topology in the small? When looked at from a distance, the surface of the ocean may appear unruffled, as a continuum; but looked at closer one may distinguish all manner of granularities, all manner of topological structures. Do the various "charges" correspond to these? Or is it that these charges are telling us of new dimensions besides the four dimensions of space and time; dimensions we have not yet apprehended? And what is the role of Planck's fundamental constant of action and Pauli's exclusion principle in this search for the meaning of charge concomitantly with the structure of space-time? We are very far yet from discovering the heart of the matter.

Roger Penrose, F.R.S.

Rouse Ball Professor of Mathematics and Fellow of Wadham College, Oxford.

Awarded the Dannie Heineman Prize of the American Physical Society and the American Institute of Physics, 1971, the Eddington Medal of the Royal Astronomical Society of London, 1975 and the Adams Prize, Cambridge University 1967.

IS NATURE COMPLEX?

1. NATURE AND NUMBER

Though Nature is undoubtedly subtle, she is surely not malicious. This, at least, we have on the authority of Einstein.[†](1) But is she complex? The word has more than one meaning: the first, nearer to "complicated" than "subtle"; and the second, a mathematical meaning concerning the nature of number. It is this second meaning that forms the main subject of this essay, though there will be relevance also to the first.

That Nature can be usefully described, at least to a considerable degree, according to the laws of number, has been in evidence for many centuries. But what is not so familiar to those without a mathematical background is that there are several *different kinds* of number, many of which are nevertheless subject to the same arithmetical laws.

Most primitive are the so-called *natural numbers*:

$$0, 1, 2, 3, 4, \ldots$$

and just about anything in our world can be quantified with their aid. We may speak of 3 apples, 17 eggs, 0 chickens, 500 people, or hydrogen atoms, or eclipses of the Moon, or lightning flashes. Next in order of abstraction comes the system *integers*:

$$\ldots, -3, -2, -1, 0, 1, 2, 3, \ldots$$

Initially these were defined as a convenience, introduced in order to make the laws of arithmetic more systematic and manageable. But there are many things in the world which cannot be quantified using them. We cannot, for example, accurately speak of -3 people in a room! On the other hand, there are various slightly more abstract ideas for which a description in terms of integers *is* appropriate, a bank balance being, perhaps, the example which springs most readily to mind. But in basic physics, also, the integers have their role to play. The clearest case of a physical attribute which seems to be accurately quantified by integers is electric charge. As far as can be told from accurate experiment, there is one basic unit of electric charge, namely that of a proton, and all other systems in nature have a charge which is an exact integer, positive, negative, or zero, when described in terms of this unit. (If the hypothetical quarks are, after all, eventually discovered existing as free particles — and intensive searches have so far failed to reveal them — then the values of this unit would have to be divided by three, but physical systems could again be described using integer charge values.) There are also other less familiar physical quantities which are described by integers, such as baryon number, the various leptonic numbers (presumably) and the quantum mechanical spin of a physical system about some axis.

While the arithmetical operation of subtraction is simplified by the introduction of negative integers, the operation of division is not. For that, the fractions are needed:

$$0, 1, \tfrac{1}{2}, -1, -\tfrac{1}{2}, \tfrac{1}{3}, \tfrac{2}{3}, 1\tfrac{1}{2}, 2, -\tfrac{1}{3}, \ldots$$

Any two such numbers can be added, subtracted, multiplied or divided (except by 0). But, while convenient mathematically, the system of fractions has seemed not to play any clear role

† "Raffiniert ist der Herrgott aber boshaft ist er nicht"; see, for example, Banesh Hoffman, *Albert Einstein, Creator and Rebel* p. 146, Hart-Davis, MacGibbon Ltd., London, 1972.

in quantifying features of nature. Instead, we must apparently leap further in abstraction and include all the so-called *real* numbers

$$0, 1, \tfrac{1}{2}, -1, \sqrt{2}, 2^{\sqrt{2}}, \pi, -6.1974302\ldots, \ \ldots$$

each of which can be represented in terms of a (signed) infinite decimal expansion. The rules of finite arithmetic are precisely the same as for fractions, but the system of reals is more complete in that certain *infinite* operations can now be performed. In particular, there are such infinite sums as:

$$\frac{\pi^2}{6} = 1 + \frac{1}{2^2} + \frac{1}{3^2} + \frac{1}{4^2} + \frac{1}{5^2} + \ldots$$

$$\sqrt{2} = 1 + \frac{1}{8}\left(\frac{2}{1}\right) + \frac{1}{8^2}\left(\frac{3\times4}{1\times2}\right) + \frac{1}{8^3}\left(\frac{4\times5\times6}{1\times2\times3}\right) + \frac{1}{8^4}\left(\frac{5\times6\times7\times8}{1\times2\times3\times4}\right) + \ldots$$

According to normally accepted views, measurements of time and distance are to be described in terms of real numbers. Indeed, it would seem that such physical notions supplied the original motivation for their invention.

The famous paradoxes of Zeno were conceived before the mathematical notion of a real number had been formally introduced. These paradoxes expressed a puzzlement about the continuous nature of time and space which had long been felt, and which have seemed to be resolved now that the rigorous mathematical concept of the real number system is at hand. In fact, such has been the success enjoyed by this number concept, that it is now hard to contemplate that time and space could be described in any other way. One gets a strong impression that time and space *are* continuous – with just the same sort of continuity that finds rigorous expression in the real number system (bearing in mind, of course, that space is three-dimensional rather than one-dimensional). But perhaps we are being blinded by appearances and by our long familiarity with the real numbers. It is this familiarity which makes *appear* natural the extrapolation to the infinitely small of a seeming continuity which is present on ordinary scales of time and space. But it could turn out that time and space do not, after all, have the kind of continuity on a very small scale that has been assigned to them. In my view – and in the views of some others[1] – such a radical change in our understanding may well be in the offing. This is a question to which I shall return later in this essay.

But if we accept the picture presented by the physics of today, the real number system provides an unending supply of elements which is called forth again and again for the quantification of physical concepts. For example, there is velocity, energy, momentum, frequency, mass, temperature, density, force, and so on. The use of real numbers for their description can, however, be traced back to the real-number continuity of space and time, so there is, in essence, nothing new to be gained from the observation that such numbers are being employed here also.

But mathematics does not stop here. Whereas subtraction, division and infinite summation have been made systematic concepts within the real numbers, the operation of solving equations has not. For example, the simple-looking equation

$$x^2 + 2 = 0$$

has no solution among the reals, whereas the apparently similar one

$$x^2 - 2 = 0$$

has the two real solutions $x = \sqrt{2}$ and $x = -\sqrt{2}$. However, for several centuries mathematicians have become used to the idea that the first equation, also, can be considered to have two solutions, written $x = \sqrt{-2}$ and $x = -\sqrt{-2}$. These are certainly not real numbers (since no negative number can have a real square root), but they are good numbers, nevertheless. Several abstractions have, in any case, had to be made in the passage from the natural numbers to the real numbers. The one further abstraction that is needed is that an extra number, denoted i, be adjoined to the system of reals, together with all those numbers formed from i by multiplying and adding real numbers to it. The number i is to be a square-root of -1:

$$i^2 = -1,$$

and the general element of the extended number system so produced, called the system of *complex numbers*, has the form

$$a + bi$$

where a and b are real numbers. In particular, the solutions to the first equation considered above are given when $a = 0$ and $b = \pm\sqrt{2}$.

At first, the notion of a complex number may seem a little bewildering. One has become accustomed to the idea that negative numbers do not "really" have square roots. It may seem, perhaps, that although these "complex numbers" are doubtless logically consistent, they serve merely as a formal device introduced for purposes of mathematical convenience. And mathematically convenient they certainly are! Not only do they enable square roots to be taken with impunity, but also cube roots, fifth roots or, indeed, any other complex power of any complex number (except perhaps zero) whatsoever. Furthermore, polynomial equations of any degree can now be solved in a completely systematic fashion. And new previously unexpected properties arise of extraordinary power and beauty. That the system of complex numbers occupies a regal place within the realm of abstract mathematical ideas, there can be little doubt. Moreover, they are not simply a "convenience"; they take on a life of their own. Yet one still may have the lingering feeling that they are not part of "reality", but merely creations of human thought.

But we must ask why it is that real numbers themselves give an impression of having this "reality" that the complex numbers seem not to have. Partly it is a question of familiarity. We are used to calculating with real numbers (or at least with finite decimal expansions) and come across them at an early stage in life. Complex numbers we encounter only much later, if at all. But more than this, it is the feeling that physical measurements, notably those of space and time, employ real numbers, whereas no such obvious physical realisation of the complex number system makes itself felt.

This, at least, seems to be the case with classical physics. With the coming of quantum mechanics the situation changes. For complex numbers play a crucial role in one of the most basic axioms of quantum theory — as will be explained shortly. But important as this fact is for the physics of the submicroscopic world, its significance begins to get lost on the macroscopic scale. Yet the role of the complex can still be discerned in the nature of phenomena at any scale

— particularly where relativity is involved. The geometry of our world is, indeed, more "complex" than it seems to be at first.

2. THE GEOMETRY OF THE COMPLEX

To try to see how this can be so, let us first indicate the standard way in which complex numbers are represented in terms of ordinary geometry. We envisage a Euclidean plane, with Cartesian axes x, y. The complex number $\zeta = x + iy$ (x, y real) is represented in this plane (called the Argand plane of ζ) by the point with coordinates (x,y). There are simple geometrical rules for the sums, products, differences and quotients of complex numbers.[2]

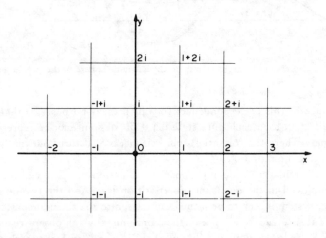

Fig. 1. The representation of complex numbers on the Argand plane.

So it seems that the world of the Flatlander could be described by the geometry of the complex number system. But what has this to do with *our* world, with its three dimensions of space and one of time? One answer is to be found in the basic rule of quantum mechanics known as *linear superposition*. If A and B represent quantum states, then for every pair of complex numbers α and β (not both zero) there exists a new state represented as $\alpha A + \beta B$, where for each distinct ratio $\alpha : \beta$ we get a physically distinct situation. Though rather formal and abstract, as stated, there is a link, apparently tenuous but in my view fundamental, between this rule and the geometry of space and time.

This may be seen at its most primitive level in the states of spin of an electron. Each such (pure) state of spin may be represented by a direction in space — where the electron is viewed as spinning in a right-handed sense about that direction. If A is the state where this direction is upwards and B where it is downwards, then any other spin direction can be represented as $\alpha A + \beta B$. There is therefore a one-to-one relation between the directions in space and the ratios $\alpha : \beta$ of pairs of complex numbers.

Now the only difference between such a ratio and just a *single* complex number is that with a ratio, *infinity* is included (i.e. when $\beta = 0$). The space of such ratios can be represented as the *Riemann sphere* S, which may be related to the Argand plane P as follows. Imagine a third axis (the z-axis) perpendicular to the plane P and through the centre 0. Take the sphere S to have unit radius and centre at 0. The north pole N of S is the place where the positive z-axis meets S. Consider a point R of P, representing the complex number ζ. The straight line RN (extended, if necessary) meets S in a unique point Q, other than N (stereographic projection). This point Q then represents the complex number ζ, on S, where ζ is now viewed as a ratio (say $\alpha{:}\beta$) and where N represents $\zeta = \infty$.

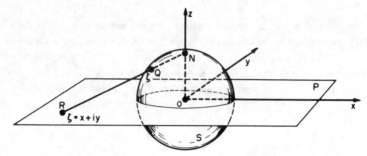

Fig. 2. Stereographic projection from the Riemann sphere to the Argand plane.

The significance of this correspondence, for our present purposes, is that if we imagine our electron situated at 0, then each pure state $\alpha A + \beta B$ of spin may be represented by a unique point of S, namely where the (oriented) spin-axis meets S. It turns out that this is precisely the point Q representing the ratio $\alpha{:}\beta$.

Thus, we have a remarkable tie-in between the directions in space and the fact that complex numbers are employed in the quantum superposition law. And the two-dimensionality of the complex number system is seen to be intimately related to the three-dimensionality of space.

This correspondence can be carried further. Imagine, now, an observer situated at the centre 0 of the sphere S. He looks momentarily out at the sky and represents each star that he sees by the point of S in that direction. In this way he can assign a complex number (ratio) ζ to each star as its coordinate. We now envisage a second observer in any state of motion whatever, who momentarily coincides with the first at the instant that the stars are observed. The second observer carries his own reference sphere S' and obtains a different complex number ζ' to label stars. The correspondence between ζ and ζ' is given by a remarkably simple formula:

$$\zeta' = \frac{\kappa\zeta + \lambda}{\mu\zeta + \nu},$$

where $\kappa, \lambda, \mu, \nu$ are complex parameters defining the relative motions and orientations of the two observers. This formula takes into account the fact that the two observers may be rotated relative to one another, the relativistic transformation between fast-moving observers and the finiteness of the speed of light being also correctly incorporated.

The formula is especially remarkable in that it is what is called *holomorphic*. This means that it is expressed analytically entirely in terms of $\zeta = x + iy$, the so-called *complex conjugate* $\bar{\zeta} = x - iy$ being nowhere involved. This holomorphic property is important and arises only because the problem has been treated correctly according to relativity.[3] No such result arises within Newtonian theory. The full power and beauty of complex analysis really only arises when holomorphic properties are considered. What we are beginning to see here is the first step of a powerful correspondence between the space-time geometry of relativity and the holomorphic geometry of complex spaces.

3. TWISTORS

We have, in effect, been just studying an aspect of the geometry of the space of *photons*. Those photons which enter the eyes of our two observers are labelled holomorphically in terms of complex numbers, as a first indication of the fact that the *space of all possible photons* is really a *complex* space. That is, the different (classical) photon states can be described holomorphically by complex parameters. Photons are examples of massless particles. All such particles travel with the speed of light, but photons are distinguished from the others by the fact that they spin about their direction of motion with a magnitude $\hbar$. For the others the spin value is some different multiple of $\hbar$. Intuitively, the massless particle may be visualised in terms of a straight line drawn in space-time, namely its world-line. But this is not strictly accurate, since the particle is not quite localised when the spin is non-zero. More correctly, the state of a (classical) massless particle should be described by its energy and momentum, and its relativistic angular momentum about some origin. It turns out that all this information can be encoded as a certain complex quantity called a *twistor*[4],[5] which is (holomorphically) parameterised by four complex numbers.

These twistors may be taken as the primary elements in an approach to physics for which space-time points are superseded in their role. Instead of describing a particle in terms of a space-time locus, we describe it using twistors. We may think of a twistor as lying, conceptually, somewhere between a point and a particle. Points themselves are to be *constructed* from the twistors. And so also are the particles. In place of the wave function of a quantum mechanical particle, it turns out we require a holomorphic function of one or more twistors. The 1-twistor particles are all massless (photons, gravitons, . . .); the 2-twistor particles appear to be "leptons" (electrons, muons, . . .); and the 3-twistor particles, "hadrons" (protons, neutrons, pions, kayons, . . .) while possibly there are also 4-twistor particles, etc., forming further families (e.g. the newly discovered J/ψ particles). Moreover, the number of twistors used can apparently be increased when needed so that an n-twistor particle can also play a role as an $(n+1)$-twistor particle. The usual space-time descriptions are accordingly dispensed with altogether, although translations to space-time terms can be achieved when needed. The description of elementary particles which arises in this way must be viewed, so far, as rather speculative. Yet the mathematics seems compelling — and the classification scheme which arises appears to be compatible with what is known from standard elementary particle theory.

My contention is, then, that the twistor description of our world is more fundamental than space-time. On a classical macroscopic scale, and when general relativity is not involved, the two descriptions are *equivalent*. But for the micro-world of quantum particles (say at the level of 10^{-13} cm or less), my claim is that the twistor description will eventually give a *more* accurate picture. And the geometry of twistors is fundamentally complex, achieving an essential interweaving of quantum-mechanical principles into the ideas of geometry.

It should be emphasised, however, that twistor theory remains somewhat conjectural and incomplete at present. Although some surprisingly economical twistor expressions are already at hand for the description of much of physics, a great deal more needs to be done. The large-scale phenomena of general relativity have yet to be adequately incorporated, though there are some very promising indications. And a twistor theory of elementary particles or of quantum fields is only just beginning to emerge. Nevertheless, the increasing role of complex spaces, as opposed to real ones, seems clear. This much arises not only from twistor theory but from many other approaches to fundamental physics. But in the twistor approach the role of these complex spaces is all-embracing. If twistor theory does indeed provide a more satisfactory framework for the description of basic physics than does the normal space-time approach, then we shall truly be able to say that nature *is* complex — in the sense of the word intended here.

4. WHERE NEXT?

What of more complicated kinds of number than the complex? There are such mathematically defined objects as quaternions, octonions and others of various different kinds. But in each case it turns out that some law of arithmetic must be sacrificed. If we wish these laws to remain intact, then complex numbers are as far as we can go. Moreover, it appears that without retaining these laws of arithmetic, a suitable function theory cannot be obtained.

It may seem ironical that complex numbers, with their own brand of two-dimensional continuity, should have had first to be brought into physics in a basic way in order to give mathematical description to various observed physical phenomena of *discreteness*. As its name implies, quantum theory is a theory designed to handle the discrete, yet it is a theory based on the use of the complex *continuum*. However, this is not so much of a paradox as it might seem at first. There is a certain kinship — even a kind of duality, reflecting the wave-particle duality of quantum physics — between complex continuity and discreteness. There is a rigidity exhibited by holomorphic functions which is quite absent in the case of functions of a real variable. Already in standard quantum theory a discreteness emerges out of the seeming morass of complex continuity. With twistor theory, it appears that we may be able to say more. In particular, the fact that electric charge occurs in integral multiples of some fixed value finds explanation within the theory — as a consequence of the rigidity of holomorphic functions.

So perhaps we shall eventually come full circle. By generalising the concept of number as far as possible away from the natural numbers of which we have direct experience, while retaining the maximum of arithmetical properties, we are apparently led back to a discreteness which we had seemed to have left behind. Perhaps, after all, the physical laws are of a discrete basically simple combinatorial nature — while at our present level of knowledge, it seems to be the complex continuum which provides the clearest route to a deeper understanding of physics.

FURTHER READING

1.   E. Schrödinger, *Science and Humanism*, Cambridge University Press, 1952.
2.   E. T. Copson, *An Introduction to the Theory of Functions of a Complex Variable*, Clarendon Press, Oxford, 1935.
3.   R. Penrose, "The apparent shape of a relativistically moving sphere". *Proc. Camb. Phil. Soc.* 55, 137-9 (1959).

Concerning twistor theory:
4.   R. Penrose, "Twistors and particles", in *Quantum Theory and the Structures of Time and Space*, edited by L. Castell, M. Drieschner and C. F. von Weizsäcker, Carl Hanser Verlag, München, 1975.
5.   R. Penrose, "Twistor theory, its aims and achievements", in *Quantum Gravity*, edited by C. J. Isham, R. Penrose and D. W. Sciama, Clarendon Press, Oxford, 1975.

Hans J. Bremermann

Professor of Mathematics and Biophysics, University of California at Berkeley.

Programmed and operated John von Neumann's pioneering computer at Princeton in 1955. Early researches into the physical limitations of computer processes influenced theoretical cybernetics and emphasized the importance of the modern theory of computational complexity.

Recent research has been concerned to some extent with complexity theory, but is mostly concerned with understanding naturally occurring "computers" and efficient methods of solving cybernetic problems.

COMPLEXITY AND TRANSCOMPUTABILITY

In this essay I wish to point out serious limitations to the ability of computers to carry out enough computations to solve certain mathematical and logical problems. The same limitations also apply to data processing by nerve nets, and thus ultimately to human thought processes.

The present era has seen advances in computing that are no less spectacular than advances in space travel. Today a single computer can do more arithmetic operations in a year, than all of mankind has done from its beginnings till 1945 when the first electronic computer became operational. I will show in the following that this achievement is insignificant when measured against the vastness of what one might call the "mathematical universe".

Computers are physical devices and as such are subject to and limited by the laws of physics. For example, no signal can travel faster than light in vacuum (3×10^8 m/sec) and this restriction applies especially to signals inside a computer. Let τ_{switch} be the switching time of computer components. (In the fastest computers τ_{switch} is of the order of 10^{-9} to 10^{-8} sec (1 to 10 nanoseconds).) The travel time of signals between different parts of a computer is determined by the distance a signal has to travel. That means τ_{travel} = distance/velocity $\geqslant$ distance/light velocity.

The travel time of signals between different parts of a computer should not exceed the switching time. Otherwise travel time would be the limiting factor in the speed of the computer.

Hence $\tau_{switch} \geqslant \tau_{travel} \geqslant$ distance/light velocity. Hence, the distance between different parts of the computer is bounded by $\tau_{switch} \times 3 \times 10^8$ m/sec $\geqslant$ distance. Thus, if τ_{switch} is 10^{-9} sec, the distances in the computer are limited to 30 cm. In other words, the entire computer must be quite small. For $\tau_{switch} = 10^{-10}$ sec the maximum size would be 3 cm, etc.

Size and signal speeds are not the only limitations. More fundamental are the following: The different components of a computer communicate with each other by signals. The reception and interpretation of a signal constitutes a physical measurement. Physical measurements are governed by the uncertainty principle of quantum mechanics. One can show: the faster the measurement the larger is the energy that is required to make the signal readable with sufficiently small error probability. If the total signalling energy is limited, then there is a trade-off between the number of distinguishable signals that can be sent and the time required to identify them. It is customary to call the logarithm (base 2) of the number of distinguishable messages the *information content* (measured in bits (binary digits)). It turns out that for given energy the amount of information that can be sent is proportional to time. The proportionality factor is given by E/h [sec^{-1}]. Where E is the energy, h is Planck's constant. *The amount of signal flow (in bits/sec) in a computer is thus limited by E/h, where E is the energy available for signalling.* (Bremermann, 1962, 1967, 1978; R. Thom, 1972, p. 143 (English translation Thom, 1975).)

This fundamental limit of data processing applies to computers, irrespective of the details of their construction. It can even be extended to computers other than digital machines (and thus becomes applicable to data processing by nerve nets). We will return to this question later.

Granted that computers are thus limited, what does it mean? Is the fundamental limit a serious barrier to computations of practical importance, or is it a subtlety without significant implications? To answer this question one must know something about the computational requirements of mathematical problems.

COMPLEXITY OF COMPUTATIONS

What exactly is the role of computation in mathematical problems? This question has been asked in earnest only in recent years. Very few results have been published prior to 1962, but since 1968 many papers have appeared that deal with this subject.

It will be useful to consider some specific examples before we attempt to examine the general question. Consider a system of two linear equations in two unknowns:

$$a_{11}x_1 + a_{12}x_2 = b_1,$$

$$a_{21}x_1 + a_{22}x_2 = b_2.$$

If $a_{11} \neq 0$ we multiply the first equation by a_{21}/a_{11} and subtract it from the second which gives

$$a_{11}x_1 + a_{12}x_2 = b_1,$$

$$(a_{22} - a_{12}a_{21}/a_{11})x_2 = b_2 - b_1 a_{21}/a_{11}.$$

Solving for x_2, substituting the result in the first equation and solving for x_1 requires altogether nine arithmetic operations.

If we have three linear equations in three unknowns we may proceed analogously. Multiply the first equation with a_{21}/a_{11} (provided $a_{11} \neq 0$) and substract it from the second. Multiply with a_{31}/a_{11} and subtract it from the third. This reduces the second and third equation to a system of two equations in two unknowns which we solve as before.

The analogous method can be applied to four, five, any number of linear equations. It is known as *Gaussian elimination* and it is an example of what mathematicians call an *algorithm*. An algorithm is a method that takes the data that come with the problem (in our case the coefficients a_{11}, a_{12} ...) and transforms them step by step until the numbers are obtained that constitute the solution of the problem (in our example the values of $x_1, \ldots x_n$ if we have n linear equations). It can be shown that the number of arithmetic operations required to carry out the Gaussian elimination algorithm is $\frac{2}{3} n^3 + \frac{3}{2} n^2 - \frac{7}{6} n$.

We may call this number the *computational cost* of the algorithm. As the number of equations, n, increases, the cost goes up, and it goes up faster than n. According to our formula the dominant term is $\frac{2}{3} n^3$, thus the computational cost (as measured in terms of arithmetic operations) increases as the third power of the number of equations.

There are other methods to solve systems of linear equations. For example, Cramer's rule which computes $x_1, \ldots x_n$ by means of determinants, and in turn there are algorithms for computing determinants. The popular algorithms of computing determinants by developing them with respect to the elements of a row or column multiplied with smaller subdeterminants have a computational cost of the order of $n! = n(n-1)(n-2) \ldots 1$. The number $n!$ increases much faster than n^3.

When we implement an algorithm on a computer we must see to it that the computational cost of an algorithm does not exceed the number of arithmetic operations that a computer can carry out within a reasonable span of time. Before the advent of electronic computers, the computational capacity of existing machines and of hand computation by humans was quite limited.

Early electronic computers could perform 100 to 1000 arithmetic operations per second. Today's computers have reached performance rates of between 10 and 100 million arithmetic operations per second, or less than 10^{13} arithmetic operations per day. If we use Gaussian elimination, we thus can at most solve $\sqrt[3]{(3 \times 10^{13})} \approx 3 \times 10^4$ linear equations in a day. (The current monetary cost of a day of computer time may run as high as £10,000.) (In practice the maximum number of linear equations that can be solved is smaller than 3×10^4 because of round-off errors that are introduced, since numbers must be limited to a fixed number of digits.)

To increase the number of linear equations that can be solved we may explore the possibility of (a) algorithms that require fewer arithmetic operations and (b) computers of greater speed. These two questions are quite different; we will first discuss question (a).

As we have seen, there is an algorithm for solving linear equations that requires of the order of $n!$ arithmetic operations. This algorithm is worse than Gaussian elimination because $n!$ grows much faster than $\frac{2}{3} n^3$. In fact, by the mid-1960s Gaussian elimination had empirically proven itself as the best all-purpose algorithm for solving linear equations. Can this be proven rigorously?

This is not an easy task. For each n we must determine the minimum of the computational costs of all possible algorithms that solve systems of n linear equations. Since the computational cost of any algorithm is positive or at most zero, the computational costs of all algorithms are bounded below by zero. Therefore, for each n the minimum exists. It is some integer between 0 and $\frac{2}{3} n^3 + \frac{3}{2} n^2 - \frac{7}{6} n$.

In the mid-sixties some mathematicians conjectured that Gaussian elimination is indeed the best algorithm for solving linear equations and tried to prove that this is the case. However, in 1968 V. Strassen (then at Berkeley, now in Zürich) described an algorithm which for large n has a lower computational cost than Gaussian elimination. Its computational cost is less than $4.7\, n^{\log_2 7}$, and $\log_2 7 \approx 2.807$. For large n this number is less than $\frac{2}{3} n^3$. Strassen did not prove that his algorithm is the best of all possible algorithms. His result provides a better upper bound for the minimal cost. It can easily be shown that any algorithm requires at least n^2 arithmetic operations (in the worst case). Thus, we have the minimal computational cost bounded by n^2 below and by $4.7 n^{2.807}$ above (Strassen, 1969).

So far we have considered a single mathematical task, namely solving systems of n linear equations in n unknowns. There are numerous other tasks of widespread interest in applications, such as finding the roots of a polynomial, solving systems of non-linear equations, optimizing a function of several variables, computing solutions of differential equations, etc. Operations research, which analyzes the efficiency of business operations and production processes, has its share of high cost computational problems. Some of these are known as "travelling salesman problem", "allocation problem", "shortest path problem", etc. It would take us too far to explain all of these problems in detail. The reader is referred to textbooks on operations research or to the paper of Karp (1972) which describes these and other problems and their interrelation in a concise way.

It may suffice to describe one of them, the travelling salesman problem, which is as follows: given $n+1$ cities $A, A_1, A_2, \ldots A_n$, any pair of cities has a distance between them. Suppose a salesman wants to visit each city exactly once, except A, the city from which he started and to which he wishes to return at the end of his trip. The problem is to find that routing of his trip which minimizes his total mileage; that is, the sum of the distances between the cities that he has visited. Each routing is given by a sequence: $A, A_{j_1}, \ldots A_{j_n}, A$ where $A_{j_i}, \ldots A_{j_n}$ is a permutation of the cities $A_1, \ldots A_n$, and A is the city where the trip starts and ends. There are $n!$ such permutations, and hence the problem can be solved by examining $n!$ mileage sums and picking the minimum. Picking the minimum of N numbers can be done with $N-1$ comparisons. Thus we could solve the problem at the cost of $n!$ comparisons between numbers (and an additional arithmetic cost of computing the mileage sums). This method, however, is not practical for large n. For example, for $n = 100$ we have (by a formula known as Stirling's formula):

$$n! = 100! > \sqrt{(200\pi)}\left(\frac{100}{e}\right)^{100}, \text{ where } e = 2.718.$$

This computational cost exceeds the capacity of any computing resource on earth.

Better algorithms than the one just described are known, but all known algorithms have a computational cost which increases faster than any finite power of n, where n is the number of cities. The problem of whether there exist algorithm whose computational cost is bounded by some polynomial in n, for all n, is unsolved. (If such an algorithm exists the problem is called *polynomial*. Whether the travelling salesman problem is polynomial is a famous unsolved problem, and lately Karp (1972) has shown that in this respect many of the problems of operations research are tied together. Either they are all polynomial or none of them is.)

The theory of the computational costs of mathematical numerical problems is known as *complexity theory*. Before 1959 it was virtually non-existent. In recent years it has made giant strides and it is developing into an entirely new branch of mathematics and theoretical computer science. For a sampling of recent results the reader is referred to Miller and Thatcher (1972). (Some further discussion of complexity problems is also contained in Bremermann (1974) and in the author's forthcoming lecture notes on biological algorithms (1978).)

For many practical mathematical, engineering and accounting tasks the computational costs of available algorithms and the capacities of available computers are satisfactory, but there are exceptions. We already mentioned operations research. Another area of difficulty is the numerical solution of large systems of differential equations, especially differential equations that are *stiff*. (That is, systems that combine processes of greatly differing speeds.) Partial differential equations pose a problem, as do the *ab initio* calculations of molecular configurations from Schrödinger's equation.

In another area, artificial intelligence, the excessive computational cost of known algorithms has been the main obstacle to having, for example, computers play perfect games of chess (or checkers, or Go). If at each move a player has k choices, then n moves comprise k^n possible move sequences. This number grows exponentially with n. For some games (like Nim) there are shortcuts that eliminate the need for searching through all the alternative move sequences. However, for chess (a game that is considered a true intellectual challenge and not mere child's play) such shortcuts have never been found. All known algorithms involve search through an exponentially growing number of alternatives and this number, when search is pursued to the end of the game, exceeds the power of any computing device.

A similar situation prevails in mathematical logic and in other branches of mathematics where there are formalized proof procedures. The search for the proof of a (conjectured) theorem always seems to involve search through an exponentially growing sequence of alternatives of formula transformations. Quite generally: most artificial intelligence problems require computing in amounts that grow exponentially with the depth of the search. The required search effort, in most cases, exceeds any available computing resource before it has reached sufficient depth to solve the problem (cf. Nilsson, 1971).

In summary: many mathematical, logical, and artificial intelligence problems cannot now be solved because the computational cost of known algorithms (and in some cases of all possible algorithms) exceeds the power of any existing computer.

THE FUNDAMENTAL LIMIT OF DATA PROCESSING

Granted that certain problems cannot be solved with existing computers, may we expect that eventually all problems will become solvable through advances in computer technology?

Computer performance has indeed increased dramatically since 1945, when ENIAC became operational. However, as I indicated in the introduction, computer performance cannot be improved indefinitely. The signal flow in a computer is limited by E/h [bits/sec], where E is the energy available for signalling. How serious is this limit?

It is easy to derive an ultimate upper bound. This bound is not meant to be realistic in the sense that it would seem practically possible to build computers that come close to this bound. Practical bounds would be much harder to derive. Thus our fundamental limit is merely a far out yardstick beyond which improvement cannot go. It is comparable to saying: astronauts cannot travel at speeds exceeding the light velocity, though in practice their speeds are much more limited. Estimates of realistic limits of the speed of space travel would be much harder to derive, having to take into consideration rocket technology, etc.

As Einstein first observed, there is an equivalence relation between mass and energy. Energy has mass, and mass, if converted, yields energy in the amount of $E = mc^2$, where c is the velocity of light in vacuum and m is the mass that is being converted. An atomic power plant is, in fact, a device for converting mass to energy.

Consider now a closed computing system, with its own power supply. Let m be the total mass of the system. This includes the mass equivalent of the energy of signals employed in the computer, while another share of the total mass is contained in the materials of which the computer and its power supply are made.

In existing computers the structural mass by far outweighs the mass equivalent of the signal energy. It would be difficult, however, to derive a realistic upper bound for this ratio. Thus, in order to avoid complicated arguments, we simply observe. *The total mass equivalent of the energy that is invested in signals cannot exceed m, where m is the total mass of the system.*

By combining this limit with the fundamental limit of the data processing we obtain:

No closed computer system, however constructed, can have an internal signal flow that exceeds mc^2/h bits per second. (Here m is the total mass of the system, c the velocity of light in vacuum and h is Planck's constant.)

This limit was derived by the author in 1961 (see Bledsoe, 1961; Bremermann, 1962). An improved argument was given by the author, Bremermann, 1967. A new discussion is to be contained in Bremermann, 1978.

The numerical value of c^2/h is 1.35×10^{47} (bits per second per gram). The number is large or small, depending upon the perspective. Existing computers process no more than $\approx 10^4$ to 10^5 bits per second per gram, and the fundamental limit appears much too large for practical purposes. When compared with the complexity of some algorithms, however, the limit appears small. The limited age and the limited size of the Universe constitute (far out) outer limits to computing. Again we choose these very unrealistic outer limits to avoid complicated arguments that could be made in order to establish more realistic and stringent bounds to the product of mass and time that could be considered as available for computing. Current estimates of the age of the physical universe run to about 20 billion years, that is 2×10^{10} years or 6.3×10^{17} sec. The total mass of the Universe is estimated as about 10^{55} gs. Thus we have 6.3×10^{72} gram seconds as an outer limit for the mass time product.

TRANSCOMPUTABLE ALGORITHMS

We call an algorithm *transcomputable* if its computational cost exceeds all bounds that govern the physical implementation of algorithms.

It can be shown that the exhaustive search algorithm for chess is transcomputable. The same is true for many algorithms of artificial intelligence and operations research. In fact, any algorithm whose computational cost grows exponentially with a size parameter n is transcomputational for all but the first few integers n.

This is a rather disturbing thought and many people have chosen to ignore it. (Analogously many people have for a long time chosen to ignore the fact that earthly resources of space, air, fossil energy and raw materials are limited.) One exception to this trend has been Ross Ashby, who more than any person in the world has emphasized the consequences of this limit in many of his writings between 1962 and his death in 1972 (cf. Ross Ashby, 1967, 1968, 1972).*

Another kind of limitation to computation is *thermodynamic*. R. Laundauer (1961) has pointed out that when in the course of computation information is discarded entropy is generated which must be dissipated as heat. How much information must be discarded when computations are carried out? Initially this question was not well understood. Recently Bennett (1973) has shown that any computation can be carried out essentially in a logically reversible way which implies that it can be done with little or no entropy generation (cf. also Landauer, 1976). In Landauer and Woo (1973) both thermodynamic and quantum limitations are discussed and an extensive bibliography is given. There are many open questions.

So far we have stated the fundamental limit for *signal flow* in a computer. Readers may wonder whether there would be an escape from the limit if we consider larger classes of computers — analogue computers, special-purpose circuitry (hardwave simulations), etc. This is not the case. In a forthcoming article I am trying to show that the limit applies to any physical implementation of any kind of algorithm. In essence, the fundamental limit is identical with the uncertainty principle of quantum mechanics.

In particular, the limit applies to nerve nets, and thus, ultimately, it imposes limits on human intelligence. This statement presupposes that the human brain is subject to the laws of physics and that it cannot solve logical and mathematical problems without implementing algorithms.

We may compare the phenomenon of transcomputability with limitations that apply to space travel. In order to reach a distant point in space the traveller has to perform a motion which requires time and energy. Since both are in limited supply the accessible portion of the Universe is limited. Analogously, in order to reach knowledge of mathematical theorems, optimal moves in a (mathematical) game, or in order to explore the trajectories of differential equations, etc., computations must be performed which require time and energy. Since both are in limited supply the accessible portion of the mathematical universe is limited.

The fundamental limit has epistemological consequences, for example the following: many systems (biological or physical) are composed of *parts*. The interactions between parts obey certain laws (e.g. gravitational interaction between mass points, electromagnetic, weak, strong interactions between elementary particles, chemical interactions between molecules, etc.). The *reductionist approach* tries to derive the total system behavior (the trajectory of the system in its state space) from the laws that govern the interactions of the component parts and from the initial state and inputs to the system. For complex systems knowledge of the systems

---

* *Note added in print*: Recently Knuth (1976) has written an article in *Science* that clearly explains the dual problems of complexity and "Ultimate Limitations" of computing (which he derives in a different way).

trajectories can be transcomputational with respect to sequential digital computation. In that case, if an analog of the system can be obtained, put in the proper initial state and if the state of the system can be observed, then the system trajectories are predictable, provided that the analog system runs faster than the original. If no such analog system is obtainable, then prediction becomes impossible, even if all the parts and the laws governing their interactions are known.

REFERENCES

Ashby, R. (1967) "The place of the brain in the natural world", *Currents in Modern Biology* 1, 95-104.
Ashby, R. (1968) "Some consequences of Bremermann's limit for information processing systems", in *Cybernetic Problems in Bionics* (Bionics Symposium, 1966), edited by H. L. Oestreicher and D. R. Moore, Gordon & Breach, New York.
Ashby, R. (1973) Editorial, *Behavioral Science* 18, 2-6.
Bennett, C. H. (1973) "Logical reversibility of computation", *IBM J. Res. Devel.* 17, 525-32.
Bledsoe, W. W. (1961) "A basic limitation of the speed of digital computers", *IRE Trans. Electr. Comp.* EC-10, 530.
Bremermann, H. J. (1962) "Part I: Limitations on data processing arising from quantum theory", in "Optimization through évolution and recombination" in *Self-organizing Systems*, edited by M. C. Yovits, G. T. Jacobi and G. D. Goldstein, Spartan Books, Washington, D.C.
Bremermann, H. J. (1967) "Quantum noise and information", *Proc. Fifth Berkeley Sympos. Math. Stat. a. Prob.*, Univ. Calif. Press, Berkeley, Cal.
Bremermann, H. J. (1974) "Complexity of automata, brains and behavior", in *Physics and Mathematics of the Nervous System*, edited by M. Conrad, W. Güttinger and M. Dal Cin, *Biomathematics Lecture Notes*, Vol. 4, Springer Verlag, Heidelberg.
Bremermann, H. J. (1978) "Evolution and Optimization", planned for publication in the *Biomathematics Lecture Notes* series, Springer Verlag, Heidelberg.
Karp, R. M. (1972) "Reducibility among combinatorial problems", in Miller and Thatcher (1972).
Landauer, R. (1961) "Irreversibility and heat generation in the computing process", *IBM J. Res. Devel.* 5, 183-91.
Landauer, R. (1976) "Fundamental limitations in the computational process", *Bericht Bunsengesellschaft Physikal. Chem.* 80 (to appear).
Landauer, R. and Woo, J. W. F. (1973) "Cooperative phenomena in data processing", in *Synergetics*, edited by H. Haken, B. G. Teubner, Stuttgart.
Miller, R. E. and Thatcher, J. W. (Eds.) (1972) *Complexity of Computer Computations*, Plenum Press, New York.
Nilsson, N. J. (1971) *Problem-solving Methods in Artificial Intelligence*, McGraw-Hill, New York.
Strassen, V. (1969) "Gaussian elimination is not optimal", *Numerische Math.* 13, 354-6.
Thom, R. (1972) *Morphogénèse et stabilité structurelle*, Benjamin, Reading, Mass.
Thom, R. (1975) *Morphogenesis and structural stability*, English translation of Thom (1972) by D. Fowler, Benjamin, Reading, Mass.
Knuth, D. E. (1976) "Mathematics and Computer Science: Coping with Finiteness", *Science* 194 . 1235-1242.

C. W. Kilmister

Professor of Mathematics at King's College, University of London.

Early interests in theoretical physics included general relativity and the work of Eddington. In recent years his interests have changed somewhat to two other fields, logic and the applications of mathematics in the social sciences. The essay in this volume represents his first considered statement in the latter field.

MATHEMATICS IN THE SOCIAL SCIENCES

1. THE PROBLEM OF THEORISING

I want to deal here with the following fundamental puzzle: mathematics has proved itself a valuable tool for carrying on sustained argument and for unifying points of view, both within its own domain and in many applications in the physical sciences. How is it that it has had no similar success in the social sciences? As a matter of fact a book appeared more than 30 years ago, which set out clearly the basic problems confronting anyone bold enough to apply mathematics in the social sciences. This was von Neumann and Morgenstern's *Theory of Games and Economic Behaviour*[1] (numbered references are to the combined list of references and notes on the text at the end of the chapter), and as I shall have cause to refer to this frequently, I denote it by vNM. Yet in the field of particular interest in vNM, economics, it is almost as if it had never been written, and the only substantial discussion of the problem of theorising – in sociology – largely misses the point of vNM. In 1943 vNM could state correctly: "Mathematics has actually been used in economic theory, perhaps even in an exaggerated manner. In any case its use has not been highly successful." The only real change today is a greater success in the provision of quantitative data, a change which obviously has far to go, so the discussion of reasons in vNM is still of interest. I can indeed make my statement of the puzzle above more precise: to what extent is the criticism of vNM on the mark, and why has it not had more effect?

von Neumann and Morgenstern dismiss the argument that mathematics is inappropriate for sciences involving the "human element" or where important factors arise which, though quantitative, defy measurement by pointing to the history of physics and chemistry and, in particular, the theory of heat. Here there was originally complete confusion between the concepts of heat and of temperature, so that theory was a prerequisite of correct measurement.

Not that von Neumann and Morgenstern would deny the inappropriateness of attempts to apply mathematics to economic problems unclearly formulated and with deficient empirical background. But the need then is to clarify the problems and improve the background. Instead, vNM utilises "only some commonplace experience concerning human behaviour which lends itself to mathematical treatment and which is of economic importance". This refers to the *theory of games* (see Section 4, p. 181), a rather simple model (to use modern jargon) of the market-place.

The value of this model is not in its empirical application, and this, I think, brings us to the important part of the argument. "Newton's creation of a rational discipline of mechanics" involved at a crucial stage the discovery of the calculus.[2] It was no straightforward application of a mathematical tool ready-made beforehand, but the development of mechanics and the calculus hand-in-hand. But, vNM argues, social phenomena are at least as complex as those of physics, so it is "to be expected – or feared – that mathematical discoveries of a stature comparable to that of calculus will be needed in order to progress in that field". The authors modestly discount the importance of the theory of games, but it does provide an example of the need for mathematical techniques not used in physics.

2. A MODEL APPLICATION

Since the call to arms in vNM has produced so little action, I shall begin by analysing it more closely. In later sections I will then review some more recent developments. To fix ideas I shall confine myself (except in Section 6) to the same field as vNM, i.e. economics. I begin the analysis by studying a very simple application which could have been made at the end of the last century. In this way I hope to concentrate on the way the mathematics is used, with all (economic) passion spent. Marshall[3] used supply and demand diagrams to determine prices in a (stable) free market. The demand diagram (D in Fig. 1) shows the amount D which would be bought for any price P measured along the bottom axis, so long as this amount is available. The supply curve, S, shows the amount S which would be brought to market when any given price is reigning. When the curves are superimposed, as in Fig. 1, they intersect at the equilibrium price P_0.

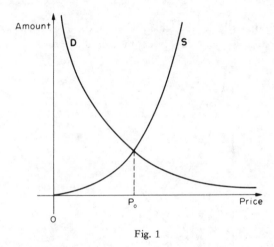

Fig. 1

Now suppose, as followers of Marshall, we try to import time into the theory so considering a non-stable system. Supply takes time to adjust to demand; let us make a simple model in which today's supply is determined by yesterday's prices, and in turn determines today's prices. And let us make the model precise by supposing that Marshall's curves are described by the equations

$$S = aP + bP^2, \qquad D = c/P, \tag{1}$$

where a,b,c are some fixed positive numbers.[4] The equilibrium price is then determined by the equation

$$bP^3 + aP^2 - c = 0.\text{[5]} \tag{2}$$

In the dynamic model, if P_n is the price on the nth day, (1) is replaced by

$$S = aP_{n-1} + bP_{n-1}^2, D = c/P_n,$$

so that, instead of (2) we have

$$\frac{c}{P_n} = aP_{n-1} + bP_{n-1}^2. \tag{3}$$

It is easier to choose new units of money so that prices are measured by x_n rather than P_n, where $P_n = x_n(c/b)^{1/3}$, which means that (3) reduces to

$$\frac{1}{x_n} = Kx_{n-1} + x_{n-1}^2, \tag{4}$$

where K stands for $(a/c)(c/b)^{2/3}$.
The equilibrium price x_0 now satisfies[6]

$$x^3 + Kx^2 - 1 = 0$$

but when we come to investigate the changes in price, Fig. 2 shows rather unwelcome conclusions. Suppose that, on any one day, prices measured in x-units are x and P on the supply

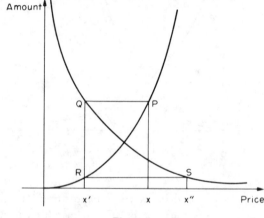

Fig. 2

curve represents the next day's supply (which is high, since prices have been taken as higher than equilibrium). Such a supply gluts the market and demand can only match supply at a price-level x' corresponding to Q on D. On the third day supply is therefore cut to R on the supply curve, producing demand T and price x'', greater than x, and so farther from the equilibrium instead, as one might hope, of approaching equilibrium. The market prices show increasingly severe fluctuations[7] as Table 1 shows (where the special case $K = 3.5$, $x_0 = 0.5$, $x = 1$ is chosen).

TABLE 1

Day	1	2	3	4	5	6	7	8	9	...
Price	1.0	0.22	1.21	0.18	1.55	0.13	2.16	0.08	3.40	...

The model is a bad one for a changing market.[8] We therefore want to know the extent to which its inadequacies can be remedied. Do they lie in the simple form (1) chosen for S and D, or are they inherent in any delayed supply-demand model? As it happens we can now say much more about the applicability of models than was possible 10 years ago and this will be dealt with in Section 7. But it will be useful first to look at possible changes. We could alter the supply curve; we might well argue that $a = 0$ (the stimulus to supply from unit change in price might be lacking altogether for very small P). More important, supply cannot continue to shoot upwards as prices increase, for limitations of "overheating" (labour shortages, raw material shortages) produce a cut-off. All in all, a supply curve like S in Fig. 3 seems more reasonable.[9]

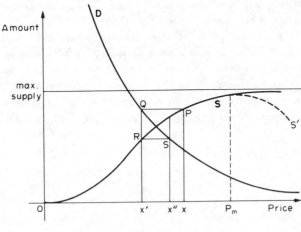

Fig. 3

And it is obvious that a suitable choice of constants can mean that the Marshall critical point gets near enough to P_m for Fig. 3 to show quite different behaviour from Fig. 2 in the dynamic case, to show, in fact, prices settling down towards the equilibrium (x'' in Fig. 3 is less than r!). Again it would be plausible to suppose that the demand curve did not fall to zero even for very high prices, but there was a constant residual "subsistence level" of demand.[10]

There is clearly no end to the variations that can be played; and each complication involves a great deal of work as soon as we get beyond the merely descriptive level. It is this fact that makes the critique of models sketched in Section 7 so important. But in any case there is a strong presumption that we are only playing at the serious task of describing reality and it is time to turn back to see what lessons are to be learnt from this analysis.

von Neumann and Morgenstern were critical, by implication, of the attempted application of calculus, the tool fashioned in the image of mechanics, to quite different phenomena. But it is surely not the calculus *per se* which is at issue; it was not used in the first model above, it proved useful in the later ones (in the proofs in the notes). Yet all models shared equally the strong air of ineffectiveness condemned in vNM. The distinction, and it is a crucial one for my argument, is between criticism of *method* and *ontology*. We do not wish to deny the usefulness of calculus as a tool for solving some particular mathematical problem, as it was used in note 10, for example, in discussing the existence or otherwise of equilibrium prices. But it is quite another matter to assume that the concepts used in calculus would be equally appropriate to incorporate into the basic formulation of an economic rather than a mechanical theory. This is what I mean by the ontological criticism.

3. THE PROBLEM OF THEORISING AGAIN

I would paraphrase an argument in physics of about 20 years ago[11] by saying that an economic theory is usefully thought of as consisting of thoughts, sentences written or spoken, observations of the economy or manipulations of it. That is, an economic theory is a union of all these things and not a pure calculating tool which exists in our minds and which we then apply to the economy. We can try to ensure that we do think in this way by substituting for the "thought" in the theory the observations of the economy which gave rise to it. This substitution is difficult, which is why it is important, because the observational thing derives meaning only by virtue of being part of the whole theory.

Let us consider from this point of view the mathematical content of vNM. Here the "thought" in the theory is the idea of a *game* of strategy, while the observational thing is that of a conflict situation between two "persons" (firms, corporations) in an exchange economy.[12] The idea of a game can be illustrated by the following simple example. Two players, P and Q, simultaneously exhibit a penny; if both show heads or both tails, P pockets both pennies, but if they show different faces, Q pockets both. From P's point of view the pay-off, in obvious notation, is

		Q's choice	
		H	T
P's	H	2	−2
choice	T	−2	2

and it is quite clear that (as the game has really no structure) it is best for each player to exhibit H or T at random. The game is fair, in the sense that, in a long run of plays, neither player will gain or lose. If, however, the rules are changed and in the new game P's pay-off is

		Q's choice	
		H	T
P's	H	2	−3
choice	T	−1	2

the new game has still an air of fairness about it. But suppose that P, instead of exhibiting H in ½ the occasions in a long run of plays, does so in p of the occasions (so $0 \leqslant p \leqslant 1$). The average pay-off to P, if Q shows H, will then be $2p - (1-p) = 3p - 1$ (since P exhibits T in $(1-p)$ of the occasions), whilst if Q exhibits T, the average pay-off to P will be $-3p + 2(1-p) = 2 - 5p$. So long as $3p - 1 < 2 - 5p$ (i.e. $p < \frac{3}{8}$) it will be best for Q (who wants to minimise the pay-off) to exhibit H; if $p > \frac{3}{8}$ he should show T. This suggests a likely hint for P; by choosing $p = \frac{3}{8}$, he puts Q in the position where he cannot gain from either H or T, and the pay-off is then $\frac{1}{8}$ to P. The full theory substantiates this surprising result.[13]

In the economy the situation is that of two persons in an exchange economy, for which the results for each depend not only on his own actions, but on those of the other. Such a problem of conflicting maxima and minima was nowhere dealt with in classical mathematics. If P,Q have more possibilities open to them (e.g. if each exhibits one face of a six-sided die) this means only that the analysis is more tedious to work out. But if more than two players are involved there are fundamental changes. Consider, for example, a particular three-person game, as devised in vNM with the intention of having no extraneous structure to confuse the argument. The three players, named 1, 2, 3, each choose the name of one of the two others. If *a* chooses *b* and *b* chooses *a*, then *a* and *b* together constitute a *couple* and they receive and share equally between them a stake (agreed before the game starts) paid by the remaining player. In such a game a couple may form, or it may not; but there is absolutely nothing more that can be said about it. So the whole structure of the game now lies outside the actual play and depends on which two players agree, by discussion beforehand, to form a coalition against the third.

If each player's agreed stake is the same the game (as far as its rules are concerned) is obviously fair, and if the stakes are different it can be made fair by giving each player a suitable (positive or negative) bonus payment for each play. But four-person games can have an intrinsic unfairness, which cannot be removed by such bonus payments.

We return to the lesson we can learn for vNM's argument in the next section.

4. THE REAL NUMBERS

We can now make better sense of the ontological aspect of vNM's argument. The mathematical objects in the theory are to have corresponding observational things. One might, for instance, make a detailed criticism (in line with the Marshallian analysis of Section 2) of the neo-classical notion of *utility*, which formed the basis of neo-classical price theory. Utility started as the property of commodities which cause them to be bought by a purchaser who seeks to enjoy their utility by consumption. In the neo-classical picture of the housewife maximising utility in filling her shopping basket it is clear that utilities are some sort of numbers. The concept lives on; it is fundamental to two-person games in vNM, and throughout they insist on a numerical estimate of advantage and disadvantage, i.e. a numerical utility for two-person games. But also vNM analyses other games, by coalitions, in terms of two-person games; though in the final sections the authors sketch a way of by-passing the two-person game so as to leave the *n*-person game in a position where numerical utilities are not required. (This sharp distinction between two-person games and others seems natural when we compare the actual methods of solution. For it is a remarkable fact that, although the two-person game exhibits new features, its solution can be reduced to a mathematical problem (linear programming) which — though admittedly not part of mathematics when vNM was written — is of a sufficiently classical flavour to turn the vNM argument applied to two-person games into a mere debating point. It is quite otherwise with other games, the theory of which is left in an incomplete form in vNM.)

I do not wish to argue against the concept of utility, only against its necessarily numerical character (a point raised before in many places), and I would want to extend this to argue against unthinking assumptions of numerical quantities in general. Such unthinking assumptions are not confined to economics, but mirror a development in mathematics itself, which is worth considering. So far we have freely used the real number field, that is, the set of numbers represented by all, terminating or non-terminating, decimals. So ubiquitous is this field that we tend to regard it as part of nature, instead of as an artefact. But, like so much else, it is part of the Greek contribution to our culture (though painstakingly reconstructed in the nineteenth century). When the Pythagoreans discovered the theorem now universally given their leader's name, it was probably by means of a similar triangle proof (Fig. 4). Triangles *ABC*, *DBA*, *DAC* are all similar, so that

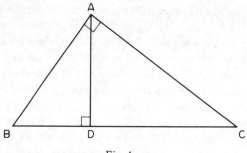

Fig. 4

$$\frac{BD}{AB} = \frac{AB}{BC}, \qquad \frac{DC}{AC} = \frac{AC}{BC}$$

and so

$$AB^2 + BC^2 = (BD + DC)\, AC = AC^2.$$

And their theory of similar triangles was, in turn, based on an argument in which two proportional sides were divided, one into p parts, on into q, so that the parts were equal. In short, the length of one side in terms of the other as unit, was p/q, a ratio of integers or *rational number*. Yet the Pythagorean theorem itself gives for the diagonal of the square of unit side $(\text{diagonal})^2 = 1 + 1 = 2$. There cannot be a fraction p/q whose square is 2, for (assuming first that p/q is in its lowest terms) this means $p^2 = 2q^2$, so p^2 is even. But the square of an odd number is odd, so p is even, $p = 2r$ (say). Hence $4r^2 = 2q^2$, $q^2 = 2r^2$ and so q is even and this is in contradiction to our assumption that p, q have no common factor.

The Greeks reconstructed their theory of magnitude — one version is in Euclid's *Elements* — and did much more. The original rational numbers formed a totally ordered field (for any two rationals r, s, exactly one of $r < s$, $r = s$, $r > s$, holds) and this ordering satisfies the axiom of Archimedes (for *any* r, s, we can take a suitable number N of copies of r so that $Nr > s$). The measure of the Greeks' achievement was the construction of the real numbers as a new, Archimedean ordered field and, as we know now, the largest (in the sense that any Archimedean-ordered field has a copy inside the reals). The uniqueness of this determination of the reals has had a great psychological effect; but, in the last decade, non-standard number systems have radically altered the picture. A non-standard system can be thought of as the reals augmented and filled in by infinite numbers and their reciprocals, the infinitesimals. The ordering can no longer be Archimedean, for the characteristic of an infinitesimal is that no number of copies of it will make it finite, but it is a total ordering. Even in the physical (geometrical) world such quantities can appear. A familiar example is the comparison of the "size" of the "horn-shaped angles" between a family of touching circles (Fig. 5). In the sense of

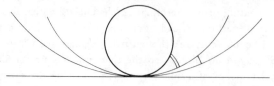

Fig. 5

the real numbers, all such angles are zero.

Now my aim here is not to promote non-standard analysis in economics, though it might just conceivably be useful in some restricted regions. (For example, the vNM argument for real numerical utilities rests on probabilities being real numbers. The axioms in vNM could be satisfied if both utilities and probabilities were non-standard; non-standard probabilities have much to commend them, for dealing with very unlikely situations.) Rather I aim to break the psychological barrier produced by the long hegemony of the real number field in mathematics, because my interpretation of the ontological aspect of the vNM *critique* is that new developments are likely to take place by use of mathematical structures different from the real numbers.

5. NEW DEVELOPMENTS – 1. AGGREGATION

It is one thing to sketch out a bold but vague scenario, as in the last sentence; it is another to carry it out. In these last three sections I wish to describe three developments which seem to be pointing in the same direction, from inside the subject, as that to which I, as armchair critic, have pointed from outside. The first of these can be seen (it was not so intended by its author) as an attempt to compromise with the ontological criticism by beginning with any amount of illicit furniture and then removing it by ingenuity and hard-work. The second (in Section 6) takes the ontological *critique* absolutely seriously and undertakes to work with only the mathematical furniture which is fully legal. Finally, in Section 7, is described a compromise which essays to show that it is possible to infringe the ontological requirements because there is a super-theory (as it were) which will pick out from the results those details which are essential and which would scarcely have been discovered without the extra props. So even without the ability to make the intellectual breakthrough of Newton and his successors, something can be done.

The first development, in economics, is in the aggregation problem, and is due to Pokropp.[14] One particular instance will serve to illustrate the aggregation problem. In the Marshallian analysis above we can ignore price, and express today's supply S directly in terms of yesterday's demand D, in suitable units, by

$$S = \frac{K}{D} + \frac{1}{D^2}. \tag{5}$$

But now imagine a two-sector economy, producing two commodities for whom supply and demand are independent, so that

$$S_1 = \frac{K_1}{D_1} + \frac{1}{D_1^2}, \quad S_2 = \frac{K_2}{D_2} + \frac{1}{D_2^2}. \tag{6}$$

The aggregation problem is simply: can one define "aggregate supply" S and "aggregate demand" D so that the macroeconomic variables S, D are related in *some* way as a result of the microeconomic relations. For example, $S_1 + S_2$ will not do as aggregate supply S, since it is obvious that S will depend on D_1, D_2 individually and not simply on some combination D of them.

Such a problem was posed by Klein[15] in 1946, and answered by Nataf 2 years later[16] within a complete calculus context.[17] The criticism of vNM was very apt, for the assumptions required exceeded any economic justification. Twenty years later, Gorman[18] weakened the assumptions to requiring only continuity, but Pokropp argues that even these are otiose, for they imply production processes with the possibility of indefinitely small variations. So Pokropp undertakes the same investigation without assuming even continuity. We need not consider his conclusions in detail; it is sufficient to point to the attempt to meet the ontological criticism *a posteriori*. The labour involved in even this amount of progress is a depressing indication of how much further progress can be expected.

6. NEW DEVELOPMENTS – 2. SIMPLICIAL COMPLEX

The second example, in which the ontological *critique* is taken absolutely seriously, is in the work of Atkin and his fellow-workers[19] in urban structure. To see how it is possible for Atkin to pursue such a rigorous path it is useful to look at some mathematical background. In this century mathematicians have learnt a valuable method of classifying surfaces, known as homology theory. The first step in this classification, whose final stage may picturesquely be described as characterising (for example) the difference between the outside surface of an (American) doughnut and that of an orange,[20] is to draw over the surface a network of triangles, somewhat like the geographers' triangulation but with this difference: the geographers are concerned with local accuracy, the triangles in homology theory need only capture the large-scale behaviour of the surface. (Similar considerations apply to more complex mathematical entities than surfaces.) The continuous surface is replaced by a finite set of triangles, some of which have edges in common, and the characterisation argument then takes place on this finite set.

Atkin reverses this motivation; if we can begin with the finite set, he argues, we need never worry about the continuous surface. Our discussion will be independent of whether it exists or not. The general finite set is a "simplicial complex" and Atkin relates this to social structure by regarding an urban community as "a collection of mathematical relations which naturally exist between physical cells and human activities". For example, he considers the set of retail trades and the set of buildings in the town centre, with the relation which identifies those trades to be found in each building.[21]

Now it is necessary to be careful not to allow the concealed assumption of "underlying surface" to be made; so, Atkin says, "an important aspect of this (and of previous) study is the acknowledgement of what constitutes 'data'. In this we are uncompromising. Data can only be the result of set-membership questions; data is yes/no observations." So his approach is to avoid "isolating" some particular problem – thought *a priori* to be able to be considered alone – because this pre-empts the question. "We do not know what depends on what until we know what depends on what." But this approach suffers certain limitations; while it is very useful in gathering hard data in the particular field of social structures, its use in economics is not at all clear. And its root and branch rejection of theorising in the usual sense of a bold hypothesis, its consequences and an attempt at their refutation implies an austerity which may limit our speculation more than it deserves.

7. NEW DEVELOPMENTS – 3. CATASTROPHES

The third straw in the wind, which takes a compromise path about the ontological criticism, arises in mathematics itself in René Thom's concept of *catastrophe*. It was foreshadowed earlier in our discussion of the Marshallian model in Section 2. Do the defects of this model arise from the specific equations chosen, or are they intrinsic? This can be answered if we know a little more about the general theory of models, and this is what catastrophe theory provides.

To illustrate the mathematics first, consider as a simple example, a quadratic equation

$$x^2 + 2bx + c = 0 ,$$

which has solutions $x = -b \pm \sqrt{(b^2 - c)}$, at least so long as $b^2 \geqslant c$ (otherwise the square root is not a real number). If $b^2 = c$ there is only one solution. Let us plot all this in a b, c plane (Fig. 6). Every point (i.e. every pair of numbers b, c) represents an equation. The points lying on the

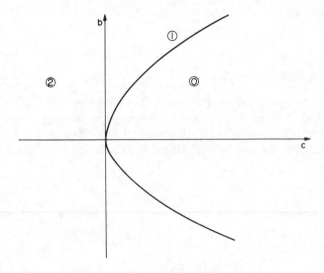

Fig. 6

parabola $b^2 = c$ are those corresponding to equations with one solution, those outside it (that is, to the left) give equations with two solutions and those to the right have no solutions. So much is elementary; Thom's insight here is topological: any point P not on the parabola is a "regular point" with the property that all nearby points (if we restrict "nearby" sufficiently) represent equations of the same general characteristics as P's. But the parabola is the set of catastrophe points, points such that any immediate neighbourhood gives equations of quite different character. Thom's great contribution is to show that, under some very general conditions, all catastrophe sets can be classified, they are all of a finite number of types, and simple examples of each are available.

We can get some idea of Thom's results by looking in more detail at the final Marshall example in Section 2 with both turn-over price P_m and residual demand, an example before which we quailed previously.[22] By suitable choice of variables, the equilibrium price can be shown to be determined by the solutions of the equation

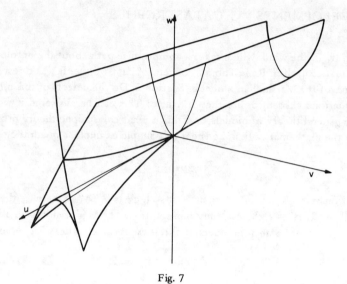

Fig. 7

$$x^4 - 6x^2 + vx + w = 0,$$

where v, w are determined by the parameters of the model (in fact by the relative sizes of response demand and residual demand, and by the shape of the supply curve). Reference to Thom's work now identifies possible discontinuities as described by the swallow-tail catastrophe (Fig. 7), where his notation used in the figure refers to

$$x^4 + ux^2 + vx + w = 0.$$

Since $u = -6$ we are concerned with only a section of the swallow's tail (Fig. 8). Again the

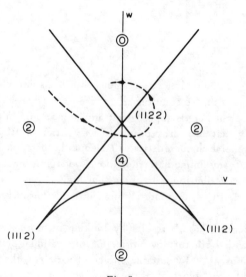

Fig. 8

curve divides the plane up according to the number of solutions (0, 2, 4) of the equation, with repeated solutions at the corners as described (1122 mean two pairs of equal solutions).

There are at least two ways of making use of this analysis. Different numerical models of the same general type correspond to different values of v, w, and so they can be compared in terms of general behaviour. The most dramatic feature is then the top wedge-shaped portion in which no equilibrium prices exist at all. This is much more serious than the former lack of stability, and arises if w is large (actually $w > 9$) and v is small. An approximate analysis shows that the restriction to the wedge corresponds to a lower limit on the response demand compared with the residual demand.[23] But instead of comparing different models we can interpret the analysis as a description of the effect of long-term trends in the market, which will be represented by slow variation of the parameters and so of v, w. Then if, as a result of outside interference (government intervention, taxation, changes in business confidence and the like) the market moves on a path like the dotted curve, unusual effects may arise whenever any edge of the curvilinear triangle is crossed. In particular most uncomfortable market conditions will arise on the later part of the path where it is impossible to fix prices which will match supply and demand.

This is only the more practical aspect of Thom's work. The more important theoretical aspect is the proof that all models of a general class involving three parameters can exhibit only one of seven elementary catastrophes, of which the swallow-tail is one. So our model is simply a particular case of all swallow-tail models; the mathematical complexity of some may be extreme but no qualitatively new phenomena will arise from these complexities. In this way, we establish an uneasy compromise with the ontological criticism; for if it requires us to alter our theories, we can conclude that the alteration will not affect the general qualitative phenomena.

Of the three indications of progress surveyed here, this last one seems to me the most promising. Much remains to be done in tackling the unanswered questions notably that of whether Thom's "general class" of models really includes all those which we want to consider in the social sciences. But it is an approach that gives promise of development, which will certainly attract many in the next decade. In this way the puzzle posed at the beginning will be answered; the reason for lack of success to date will be seen to be the lack of mathematical tools specifically designed for the purpose, and the lack of effect of vNM will be seen as caused by their failure to distinguish method from ontology.

REFERENCES AND NOTES

1. J. von Neumann and O. Morgenstern, *The Theory of Games and Economic Behaviour*, Princeton, 1944.
2. Over both mechanics and the calculus, vNM may be slightly over-generous to Newton. This does not affect the argument at all.
3. Alfred Marshall, 1842-1924, the great English economist and teacher of Keynes.
4. The coefficients a, b, c have ready economic interpretations. Since $c = PD$ it represents demand measured in money terms; so this money-measured demand is assumed constant, which might perhaps arise by a constant money-supply and constant velocity of circulation. Next, a gives the slope of the S, P curve for $P = 0$, a measure of the response of producers to price changes for low P, i.e. marginal supply-price ratio at low prices. (This might well be thought zero, or nearly so, in a plausible model.) Lastly, $2b$ is the rate of change of the marginal supply-price ratio with price, a measure of the "pick-up" in the market as prices rise.
5. It is easily calculated that this is a system in which the economist would recognise a price elasticity of supply e, given by

$$e = 1 + [1 + (a/bP)]^{-1}$$

and a price elasticity of demand of unity.

6. For different K we have different values of e in note 5; a table of solutions of cubic equations enables one easily to calculate that the elasticities at equilibrium are

K	e
0	2
0.16	1.87
1.59	1.44
7.94	1.29
15.87	1.28
Very large	1.28

7. This is the economists' "cobweb" result; the general conclusion being well known to be that, for stability, the slope of the demand curve at x_0 must be greater than that of the supply. Putting $D = dc^{2/3} b^{1/3}$, $S = sc^{2/3} b^{1/3}$, we have

$$d = 1/x, \quad s = Kx + x^2 ,$$

and the cobweb condition becomes $1/x_0^2 > K + 2x_0$, i.e. $1 > Kx_0^2 + 2x_0^3$. Since, however, $1 = Kx_0^2 + x_0^3$, this condition can never be satisfied.

8. More positively, this model shows the impossibility of one price-determining mechanism; such a use of models is commonplace now in economics. A striking example is P. Sraffa, *The Production of Commodities by Means of Commodities*, Cambridge, 1960, where an elegant model of production as a circular process does not pretend a close fit with the real economy, but is used to show that even with favourable assumptions like a uniform rate of profit it is impossible that rate of profit should be determined by marginal productivity of capital.

9. A simple formula for the curved S is not available, but for the dotted curve S' we may surmise

$$S' = bP^2 - dP^3 ,$$

where d is a suitable positive constant, and S, S' will agree so long as the prices involved are significantly less than P_m, where P_m gives $dS'/dP = 0$, i.e. $P_m = 2b/3d$. (This defines d as $2b/3P_m$ where P_m is "turn-over price".) The equation to determine equilibrium prices is now

$$f(P) \equiv dP^4 - bP^3 + c = 0$$

and $f(P)$ has a single minimum value at $P = 3b/4d$. So there are two solutions so long as $f(3b/4d) < 0$, i.e. $cd^3 < 81b^4/256$. But one such solution must be less than P_m, so $f(P_m) < 0$, which strengthens the condition to $cd^3 < 8b^4/81$. Now the cobweb result requires (it is easily seen) $dP^4 > c$ at the equilibrium value, so $f((c/d)^{1/4}) > 0$, which implies that $cd^3 > b^4/16$. In all, then, the new model has stability of prices so long as the one numerical parameter $A = cd^3/b^4$ satisfies $0.0625 < A < 0.0988$.

10. If e is the residual demand, we then have

$$D = \frac{c}{P} + e, \quad S = bP^2 - dP^3$$

so that the equilibrium is determined by

$$f(P) \equiv dP^4 - bP^3 + eP + c = 0 .$$

To what extent will e make any significant difference? Certainly the one numerical parameter A will no longer suffice to determine behaviour. Another arises as follows:

$$f'(P) = 4dP^3 - 3bP^2 + e, f''(P) = 12dP^2 - 6bP$$

and so f' has turning values at $P = 0, b/2d$ and so f' has maximum value e and minimum $b^3/d^2 (B - \frac{1}{4})$ where $B = ed^2/b^3$. Hence f' has two positive zeros if and only if $B < \frac{1}{4}$ and this in turn may be used to investigate whether f has a unique positive zero less than P_m. Are A and B sufficient? (They should be, of course, since we can always remove the coefficient b by adding a suitable constant to the roots, and dividing by d then gives an equation with only two parameters.)

11. E. W. Bastin and C. W. Kilmister, "The concept of order, I. The space-time structure". *Proc. Camb. Phil. Soc.* 50, 178 (1954).

12. The idea of a mathematical theory of games of strategy seems to be due to E. Borel in 1923, but von Neumann proved the fundamental theorem (the existence of an optimal mixed strategy for any

two-person zero-sum game) in 1928.

13. In fact, the average pay-off will be

$$2pq + 2(1-p)(1-q) - 3p(1-q) - q(1-p)$$

$$= 8(p - \tfrac{3}{8})(q - \tfrac{5}{8}) + \tfrac{1}{8}.$$

The optimal strategies for P, Q are $(H, T) = (\tfrac{3}{8}, \tfrac{5}{8})$ or $(\tfrac{5}{8}, \tfrac{3}{8})$ respectively.

14. F. Pokropp, *Aggregation von Produktionsfunktionen*, Springer, Berlin, 1972.

15. L. R. Klein, *Econometrica* **14**, 303 (1946).

16. A. Nataf, *Econometrica* **16**, 232 (1948).

17. More explicitly, all the functions assumed in the investigation were differentiable, and had positive partial derivatives.

18. W. M. Gorman, *Rev. Econ. Studies* **35**, 367 (1968).

19. Very full details are given in the series of research reports to the Social Science Research Council on the *Urban Structure Research Project* (University of Essex, 1972, 1973 (two reports), 1974) and the continuation *Methodology of Q-analysis* in 1975. But see also: R. H. Atkin, *Mathematical Structure in Human Affairs*, Heinemann, 1974; R. H. Atkin, *Environment and Planning* B, **I**, 51 (1974) and **II**, 173 (1974).

20. D. B. Scott (a verbal communication).

21. The abstract definition of a simplicial complex, in terms of two finite sets X, Y and a relation λ between them, is that the complex $K_Y(X, \lambda)$ is (i) a collection of simplices σ_p, where (ii) each σ_p is defined by being a subset of $(p + 1)$ elements of X for which there exists a y in the relation λ to all of them and (iii) the σ_0's being identified with the members of X.

22. The equation for equilibrium prices in note 10 above can be rewritten, if $P - b/4d = y/4d$, as

$$y^4 + py^2 + qy + r = 0,$$

where

$$p = -6b^2, q = 8(8ed^2 - b^3),$$

$$r = 64bed^2 + 256cd^3 - 3b^4.$$

Putting $y = bx$ and using the parameters A, B of notes 9, 10 we get

$$x^4 - 6x^2 + vx + w = 0,$$

where $v = 8(8B - 1)$, $w = 64B + 256A - 3$.

23. The corners of the triangle are easily seen to be at $(0, 9)$ corresponding to $x^4 - 6x^2 + 9 \equiv (x^2 - 3)^2$, and at $(\pm 8, -3)$ corresponding to $(x + 3)(x - 1)^3$ or $(x - 3)(x + 1)^3$, respectively. The *fairly* straight sides of the triangle can therefore be approximated to by straight lines $w = \pm \tfrac{3}{2} v + 9$. The wedge-shaped region is then approximately

$$w > \tfrac{3}{2} v + 9, \quad w > -\tfrac{3}{2} v + 9.$$

It is simplest to deal with the two cases separately. If $B > \tfrac{1}{8}$, so that $v > 0$, the restriction is $w > 9 + 3v/2$, i.e. $A > \tfrac{1}{8} B$. If $B < \tfrac{1}{8}$, so that $v < 0$, then $w > 9 - 3v/2$, $A > \tfrac{3}{32} - \tfrac{5}{8} B$. If we take P_m as unit of money, so that $P_m = 1$ and $b/d = \tfrac{3}{2}$, then

$$A = (\tfrac{2}{3})^3 (c/b), \quad B = (\tfrac{2}{3})^2 (e/b)$$

and so A, B are measures of response demand and residual demand in terms of the "pick-up" in supply b.

Heini Halberstam

Professor of Pure Mathematics, University of Nottingham.

Member of the Council of the London Mathematical Society and
Vice-President 1962-3, 1974-7. Vice-President of the Council of the
Institute of Mathematics and its Applications 1972-4, Member of the
U.K. National Committee for Mathematics.

SOME UNSOLVED PROBLEMS IN
HIGHER ARITHMETIC

> *... a mathematical problem should be difficult in order to entice us, yet not completely inaccessible, lest it mock our efforts. It should be to us a guidepost on the tortuous paths to hidden truths, ultimately rewarding us by the pleasure in the successful solution.*
>
> D. Hilbert

1. Number theory, or the higher arithmetic, is one of the oldest branches of mathematics. The bland appearance of the sequence of integers is profoundly deceptive; it masks an inexhaustible multitude of beautiful patterns and tantalising questions which have fascinated scholars and amateurs alike since earliest times. Usually such problem situations emerge from experimentation, and are easy to describe; but to account for them is apt to be hard and often extremely difficult. It is common nowadays to assess the development of a branch of mathematics by the extent to which its primary matter has been incorporated in a general and comprehensive intellectual edifice; but while many early arithmetical investigations have indeed grown into imposing theories, the basic appeal of number theory is still, and always will be, that most questions do not fit readily into an existing theory. Knowledge and experience are important, of course; but it is the opportunity for *ad hoc* innovation, for the new idea or fresh insight, for novel connections, that draws inquiring minds to the mysteries of number. In this respect (as also in the importance attached to experiment) the higher arithmetic is close to the natural and biological sciences: we know a great deal, but the extent of our ignorance is limitless.

In view of what I have said, it is hardly surprising that there already exists a literature concerned with charting the unknown in the theory of numbers; and the interested reader should consult[a] the articles of Erdös[1,2] and the books of Shanks,[3] Sierpinski[4] and Mordell[5] for further reading. Indeed, within the space available I can do little more than present a selection from these works which reflects my own interests.

2. The prime numbers are the bricks from which all the positive integers (other than 1) are constructed by multiplication. Their supply is infinite, as was already known to Euclid; for if $p_1 = 2, p_2 = 3, \ldots ,p_n$ are the first n primes, the integer $p_1 p_2 \ldots p_n + 1$ must be divisible by a prime other than $p_1 p_2, \ldots,$ or p_n. It follows that p_{n+1} exists; moreover, it satisfies

$$p_n < p_{n+1} \leqslant p_1 \ldots p_n + 1. \tag{2.1}$$

Observe that 2+1 = 3, 2.3+1 = 7, 2.3.5+1 = 31, 2.3.5.7+1 = 2011, 2.3.4.7.11+1 = 2311 are all primes! But this "pattern" breaks down eventually; indeed, it is not known even whether $p_1 \ldots p_n + 1$ is prime infinitely often, nor whether it is infinitely often composite. As every school child knows, and Gauss was the first to prove, every integer $m > 1$ has a unique "canonical" representation[b]

$$m = p_1^{\alpha_1} \ldots p_r^{\alpha_r} \qquad (p_1 > \ldots > p_r) \tag{2.2}$$

as a product of primes, where $\alpha_1, \ldots ,\alpha_r$ are positive integers. The proof of this basic result (the so-called Fundamental Theorem of Arithmetic) is elementary but delicate; it is significant that the proof depends on both the multiplicative *and* additive structure of the integers. An interesting way of stating the theorem is that if $p_1, \ldots ,p_r$ are distinct primes then

$\log p_1, \ldots, \log p_r$ are *linearly independent* over the rationals in the following sense:

$$\text{if } \lambda_1 \log p_1 + \ldots + \lambda_r \log p_r = 0 \quad (\lambda_1, \ldots, \lambda_r \text{ rational})$$

then $\lambda_1 = \ldots = \lambda_r = 0$. It follows at once, for example, that $\log 3/\log 2$ is an irrational number. If, on the other hand, $\lambda_1, \ldots, \lambda_r$ are not all zero, then the *linear form*

$$\lambda_1 \log p_1 + \ldots + \lambda_r \log p_r$$

cannot be too small (in absolute value). To obtain good lower bounds for such forms (and indeed for more general linear forms involving algebraic numbers) is the subject of lively current research, spearheaded in recent years by Professor A. Baker of Cambridge, and has many important arithmetical applications.[c] Some of these will be mentioned later, but let me describe a particularly striking instance right away.

In 1894 Catalan enunciated the conjecture that the equation

$$x^m - y^n = 1$$

in integers $x > 1$, $y > 1$, $m > 1$, $n > 1$ has no solutions other than $m = y = 2$, $n = x = 3$. Last year R. Tijdeman succeeded in showing that, at any rate, this equation has at most a finite number of solutions, by applying refinements of Baker's famous method in a surprisingly simple way. A striking aspect of the method is that it gives explicit bounds for possible solutions, so that, in principle, all solutions may be identified by a computer search. Unfortunately, the bounds are so huge that no existing computer is sufficiently powerful. The corresponding question for the equation

$$ax^m - by^n = c \quad (a,b,c \text{ constant})$$

is still open, although Y. V. Čudnovsky in Russia claims to have dealt with the case $a = b = 1$.

The number $\tau(m)$ of all positive integer divisors of m is (cf.(2.2)) given by $\tau(m) = (\alpha_1 + 1) \ldots (\alpha_r + 1)$. Clearly $\tau(m) = 1$ if and only if $m = 1$ and $\tau(m) = 2$ if and only if m is prime. As one might expect, writing down the values of $\tau(m)$ as m runs successively through the positive integers gives no impression of regularity; nevertheless, we meet here the statistical phenomenon that $\tau(m)$ does behave well on *average*; in fact, it behaves like $\log_e m$, in the sense that, as Dirichlet proved long ago,[d]

$$T(x) = \sum_{m \leqslant x} \tau(m) = x \log x + (2\gamma - 1)x + 0(x^\delta) \quad (x \to \infty)$$

with $\delta = \frac{1}{2}$. But this beautiful mean value formula is still the subject of research, for it is known to hold with values of $\delta < \frac{1}{2}$ and even with $\delta < \frac{1}{3}$; on the other hand, we know from a limitation principle of G. H. Hardy, that $\delta < \frac{1}{4}$ is impossible. The gap between $\frac{1}{3}$ and $\frac{1}{4}$ seems small but has proved formidably hard to bridge, despite the beguiling apparent simplicity of the problem; for $T(x)$ is just the number of points with integer coordinates in the shaded hyperbolic region (Fig. 1). Put in this way, we see that this problem is one of a whole class of similar questions, all characterised by the same difficulty. The no less fascinating companion problem involving a circular boundary is at about the same stage of development.

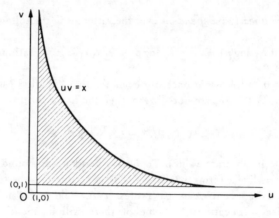

Fig. 1.

Before leaving $\tau(m)$, let us mention an old conjecture of Erdös, that there exist infinitely many values of m for which $\tau(m) = \tau(m+1)$. This would follow at once if we could prove that, for infinitely many primes p, $2p+1$ is a product of two distinct primes (for example, $2.7 + 1 = 15 = 3.5$), since then $\tau(2p) = 4 = \tau(2p+1)$; but this latter question is profoundly difficult (see paragraph 5) and will be discussed later in its proper context.

One of the oldest, and still one of the hardest, problems about primes also goes back to Euclid and derives from another divisibility question. Let $\sigma(m)$ denote the sum of all positive integer divisors of m. Following the Greeks, we say that m is *perfect* if $\tau(m) = 2m$, *deficient* if $\tau(m) < 2m$ and *abundant* if $\tau(m) \geqslant 2m$ — thus 6 is perfect ($1+2+3+6 = 2.6$), as is 28, any prime power (e.g. 8) is deficient, and any number of the form $2^{\alpha-1}(2^\alpha-1)$|is abundant ($\alpha = 2,3, \ldots$). This classification of the integers was regarded as significant in numerology well into medieval times (one might encounter statements like "the Creation was perfect because God created the world in *six* days" or "the second creation of the world is descended from the *eight* souls in Noah's ark and is therefore inferior to the first", eight being a deficient number, etc.); it remains of interest because the underlying arithmetical questions are extraordinarily difficult. We know from Euler's completion of Euclid's result that an *even* m is perfect if m is of the form

$$2^{\alpha-1}(2^\alpha-1) \ where \ 2^\alpha-1 \ is \ prime.$$

Hence the set of perfect numbers is infinite if there exist infinitely many primes of the form $2^\alpha - 1$; we conjecture that the latter is true, but have no idea how to set about proving it. It is easy to see that[e] $M_\alpha = 2^\alpha - 1$ cannot be a prime unless α is one too; and in fact M_2, M_3, M_5 and M_7 are prime numbers. However, M_{11} is composite and it is known that the only other primes $p \leqslant 257$ for which M_p is prime are 13, 17, 19, 31, 61, 89, 107 and 127. (M_{4423} is a prime; until about 10 years ago it was the largest known prime number!) I mention these numerical facts to indicate how difficult it is to factorize large numbers, even ones having an apparently simple structure. It is plausible that the immediate neighbours of highly composite numbers — that is, numbers like M_p, or the Fermat numbers $2^{2^n} + 1$, or the Cullen numbers $n2^n + 1$, or $n! + 1$ or numbers of the form $2^{2^{\cdots^2}} + 1$ — have relatively few prime factors, and one might expect therefore to be able to show at least that, if $P(m)$ denotes the largest prime factor of m, then $P(M_\alpha)$ is large as a function of α. However, even this question is exceptionally difficult; while it is known that $P(M_\alpha) > \alpha$ for every positive integer α, proving that $P(M_\alpha) > 2\alpha$ (for $\alpha >$

12) is already rather complicated. We suspect that

$$\lim_{\alpha \to \infty} P(M_\alpha)/\alpha = \infty,$$

and as a pointer in this direction we have the recent result of Stewart that there exists an integer sequence $\alpha_1 < \alpha_2 < \ldots$ on which $P(M_{\alpha_n})/\alpha_n \to \infty$ as $n \to \infty$; moreover, for all sufficiently large primes p, there is the sharper lower estimate

$$P(M_p) > \frac{1}{2} p \, (\log p)^{\frac{1}{4}}.$$

Stewart has also proved recently (using lower estimates for linear forms in logarithms mentioned above in all these results) that, for all positive integers n,

$$P(2^{2^n}+1) > cn2^n,$$

with $c > 0$ some absolute constant. Erdös and Stewart have shown also, by elementary means this time, that, for all positive integers n,

$$P(n!+1) > n + (1-\epsilon_n) \log n/\log \log n$$

where $\epsilon_n \to 0$ as $n \to \infty$; and that, for infinitely many positive integers n,

$$P(p_1 \ldots p_n+1) > p_{n+k}, \quad \text{where } k > c \log n/\log \log n$$

with $c > 0$ some absolute constant.

Returning to perfect numbers, it is conjectured that there exist no *odd* numbers of this kind. Experimental evidence shows that even among abundant numbers odd integers are relatively rare (945 is the first odd abundant number), and it has been shown that if an odd perfect number exists it must be exceedingly large (for example, it must be greater than $e10^{36}$ and have at least 2700 distinct prime factors). It appears that this ancient question is still exceptionally intractable.

3. The best known divisibility theorem of elementary number theory is Fermat's congruence: For every integer a, $a^p - a$ is divisible by p; or, as we may write using the language of congruences,

$$a^p \equiv a \bmod p. \tag{3.1}$$

For a long time it was believed that the converse is true also — in other words, that if m is an integer such that

$$a^m \equiv a \bmod m \tag{3.2}$$

for every integer a, then m is prime. This turns out to be false, as we can check in the case of $m = 561 = 3.11.17$, and raises the question whether there are many, perhaps infinitely many, *composite* m for which (3.2) holds for all a. Such numbers m are called *Carmichael numbers* —

5.29.73, 7.31.73, 5.17.29.113.337.673.2689 are other examples — and we do not know the answer.

For a fixed integer $a > 1$, however, we can show without much difficulty that

$$m = \frac{a^{2p}-1}{a^2-1} = \left(\frac{a^p-1}{a-1}\right)\left(\frac{a^p+1}{a+1}\right) \qquad (3.3)$$

satisfies (3.2) for each odd prime p not dividing $a(a^2-1)$, so that there exist infinitely many composite numbers m satisfying (3.2). In the special case of $a = 2$ we refer to all such numbers m as *pseudo-primes*; from (3.3) we see that all integers of the form $\frac{1}{3}(4^p-1), p = 5,7,11, \ldots,$ are examples of *odd* pseudo-primes, and the least of these (corresponding to $p = 5$), 341, is actually the first pseudo-prime because it can be checked that if $n \leqslant 340$ and 2^n-2 is divisible by n, then n is prime. Each of the numbers $m = \frac{1}{3}(4^p-1)$ $(p \geqslant 5)$ can itself generate an infinite class of odd pseudo-primes, for one can show that if m is an odd pseudo-prime then so is 2^m-1. It was not until 1950 that D. H. Lehmer discovered an *even* pseudo-prime, 161,038, but in the following year Beeger showed that, in fact, there exist infinitely many such numbers. Nevertheless, it is unlikely that all classes of pseudo-primes have been identified yet, and although we do know that they are relatively rare we have still to determine finer details about the way in which they are distributed.

Returning to (3.1) in the case $a = 2$, we now write it in the form

$$2^{p-1} \equiv 1 \bmod p \qquad \text{for every prime } p > 2,$$

and we ask: do there exist primes p for which

$$2^{p-1} \equiv 1 \bmod p^2 ? \qquad (3.4)$$

Two examples are $p = 1093$ and $p = 3511$, and we know there are no others below 100,000. It may be that there are only finitely many solutions of (3.4), but we do not know. The question is of considerable interest in connection with one of the most famous unsolved problems in the whole of mathematics, Fermat's conjecture that if $n \geqslant 3$ the equation

$$x^n + y^n = z^n$$

has no solution in positive integers x,y,z. (In the case $n = 2$ there are infinitely many solutions — the so-called Pythagorean triples, of which 3,4,5 and 5,12,13 are well-known illustrations.) Although formidable tools have been developed to attack Fermat's conjecture[f] — much of modern algebra derives from such attempts — we still do not know even that there exist infinitely many n for which the conjecture is true. The case $n = 4$ can be disposed of in elementary fashion, by Fermat's own highly effective method of "infinite descent"; but the case $n = 3$, which also has been settled, requires in addition quite sophisticated algebraic ideas. In view of this, we may clearly suppose, in discussing Fermat's equation, that n is an odd prime $p \geqslant 5$. The so-called "first case" of Fermat's conjecture refers to the equation

$$x^p + y^p = z^p \qquad (p \geqslant 3) \qquad (3.5)$$

in which x,y,z are positive integers, and none of x,y,z is divisible by p. It was proved by

Wieferich and Miramanoff that if a solution of this kind exists, then (3.4) must hold and also

$$3^{p-1} \equiv 1 \bmod p^2.$$

All experimental evidence suggests that if such primes exist they must be exceptionally rare. We might mention here also the criterion of Vandiver: if (3.5) has a solution in positive integers, then we may clearly suppose that such a solution triple x,y,z exists with highest common factor 1; and in that case

$$x^p \equiv x \bmod p^3, \quad y^p \equiv y \bmod p^3, \quad z^p \equiv z \bmod p^3.$$

We end this discussion of Fermat's congruence with one more open question. It is easy to deduce from (3.1) that

$$1^{p-1} + 2^{p-1} + \ldots + (p-1)^p + 1 \equiv 0 \bmod p.$$

It has been conjectured, and checked for all integers up to 10^{1000}, that this divisibility characterizes primes only; but we have no idea how to set about a proof.

4. Consideration of even perfect numbers led us to ask whether there are infinitely many primes in the sequence $2^{\alpha} - 1$ ($\alpha = 2, 3, \ldots$), and we saw that this is a very deep question. In fact, the same question is difficult in relation to any sequence of integers (not excluded by trivial considerations). Let us begin with the simplest of these, the sequence $2, 3, 4 \ldots$ of natural numbers > 1. If $\pi(x)$ denotes the number of primes not exceeding x, we do know, of course (see paragraph 2), that $\pi(x) \to \infty$ as $x \to \infty$, but many difficult questions about the distribution of primes present themselves.

For a start, one might ask how rapidly $\pi(x)$ tends to infinity with x. By systematic use of empirical evidence Gauss and Legendre were led at the end of the eighteenth century (independently of each other) to the conjecture that $\pi(x)$ behaves, for x large, like $x/\log x$; and a hundred years later (in 1896) Hadamard and de la Valleé Poussin (also independently of each other) confirmed this by proving that

$$\lim_{x \to \infty} \frac{\pi(x)}{x/\log x} = 1. \tag{4.1}$$

Gauss (still in his teens!) conjectured even that

$$\mathrm{li}\, x = \int_2^x \frac{dt}{\log t}$$

is a better approximation to $\pi(x)$ than $x/\log x$, and here too he was right; writing

$$E(x) = \pi(x) - \mathrm{li}\, x,$$

we know that $E(x)$ is at most of the order of magnitude of

$$x e^{-c\sqrt{\log x}} \quad (c \text{ a positive constant}) \tag{4.2}$$

and with much hard work one can do even a little better. However, the true order of magnitude is still unknown; it cannot be much smaller than $\sqrt{x}$ — that we know from the pioneering work of Littlewood — and we expect this to be close to the truth. But the gap between (4.2) and $\sqrt{x}$ is enormous.

The asymptotic formula (4.1) is known as the Prime Number Theorem, and we may regard it as a global statement about the *density* of primes. It tells us little about the finer distribution of primes among the integers — for instance, how often are consecutive primes close to one another, and how far apart at worst can they be? These are questions of tantalising difficulty. We think that prime pairs such as 5 and 7, 11 and 13, 17 and 19, 29 and 31 occur frequently — about $x/(\log x)^2$ times between 1 and x (x large) — but so far we have been unable to prove even that these prime "twins" occur infinitely often. On the other hand, while there are certainly arbitrarily long runs of consecutive composite integers (just take the succession, $m! + 2, m! + 3, \ldots, m! + m$ for any integer $m \geqslant 2$) we believe that for every $m > 1$ there is a prime between m^2 and $m^2 + m$; but we cannot prove it. The best result (due to M. N. Huxley) known at present is that there is a prime between m^2 and $m^2 + m^\theta$ for any number $\theta > 7/6$ provided $m \geqslant m_0(\theta)$ (where $m_0(\theta)$ is a sufficiently large number depending on θ). We believe, on the basis of a statistical argument of Cramer, that if p_n, p_{n+1} are successive primes, there exists an infinite sequence of integers n on which

$$\lim_{n \to \infty} \frac{p_{n+1} - p_n}{(\log n)^2} = 1,$$

and that, given any $\epsilon > 0$,

$$\frac{p_{n+1} - p_n}{(\log n)^2} < 1 + \epsilon$$

for all $n \geqslant n_0(\epsilon)$.

What we know in the other direction (from the work of Rankin) is that for infinitely many n

$$p_{n+1} - p_n > c(\log n)(\ln_6 n)/(\ln_3 n)^2$$

where c is a positive number a little smaller than $1/3$ (and $\ln_r n$ is here the r-fold iteration of log, so that $\ln_3 n = \log \log \log n$). On the basis of other work of Cramer, Erdös has conjectured that

$$\lim_{N \to \infty} \frac{1}{N(\log N)^2} \sum_{n=1}^{N} (p_{n+1} - p_n)^2$$

exists and is positive, but this seems hopelessly out of reach at the present time[g]. One could go on conjecturing along these lines indefinitely, but perhaps enough has been said to indicate how far we are from understanding fully the way in which the primes are distributed among the integers.

5. Let us, nevertheless, venture into still deeper water: how are the primes distributed relative to integer sequences other than the sequence of all the natural numbers; recall that we have already raised this question with respect to the sequence $2^n - 1 (n = 1, 2, \ldots)$? The next simplest case, and from the point of view of applications perhaps the most important, is an arithmetic progression, say

$$\ell, \ell + k, \quad \ell + 2k, \quad \ell + 3k, \ldots;$$

we require that HCF $(\ell, k) = 1$, since otherwise all terms except possibly the first are composite. We know from Dirichlet's seminal memoir of 1837 that such a progression contains infinitely many primes, and we know from later work of de la Vallee Poussin even that each of the progressions with common difference k contains roughly the correct proportion[h] $1/\phi(k)$ of all primes. More precisely, if $\pi(x;k,\ell)$ denotes for a given k the number of primes $p \leqslant x$, $p \equiv \ell \bmod k$, then the difference

$$E(x;k,\ell) = \pi(x;k,\ell) - \frac{1}{\phi(k)} \operatorname{li} x$$

is certainly no larger in order of magnitude than the expression (4.2) which we met in connection with the Prime Number Theorem, and again, the true order of magnitude is expected to be much smaller. However, for application we need to know about the distribution of primes, as measured by $E(x;k,\ell)$, relative to not just one fixed progression but to *all* progressions with common differences $k < x$, and here we run into serious difficulty. All we have at present is information of an *average* kind: for all k almost up to $\sqrt{x}$, $|E(x;k,\ell)|$ is on average no larger than $\sqrt{x}$ in order of magnitude (Bombieri's theorem); and for all k almost up to x, $|E(x;k,\ell)|$ is on average, but only in the "mean square" sense, like $\sqrt{(x/k)}$. It is fortunate that for some very important applications such average results, especially the theorem of Bombieri, have proved sufficient. There are all kinds of other questions one might raise in connection with $\pi(x;k,\ell)$ and $E(x;k,\ell)$. For example, given k, how large does x have to be to ensure that $\pi(x,k,\ell) > 0$; in other words, how large is the *least* prime $P(k,\ell)$ in the arithmetic progression $\ell \bmod k$? We believe that $P(k,\ell)$ cannot be much larger than $k \log k$ and we know this is true in a certain average sense. According to a famous result of Linnik, $P(k,\ell) < k^C$ for some positive constant C, and considerable effort has gone into determining admissible values of C. The present "record" is $C \leqslant 80$; this is due to M. Jutila, but recently S. Graham has shown that $C = 36$ is admissible. Or we may ask for comparisons between $\pi(x;k,\ell_1)$ and $\pi(x;k,\ell_2)$. Littlewood proved, contrary to impressive numerical evidence, that $\pi(x;4,3) - \pi(x;4,1)$ changes sign infinitely often, but we have, so far, no unconditional results of a similar kind for general k, despite the impressive studies of Turán and Knapowski in so-called "comparative prime number theory". Absolutely nothing is known about the fluctuations in $|E(x;k,\ell)|$ for a given k as ℓ ranges over values between 1 and k coprime with k, and even numerical evidence about this would be extremely hard to compile.

The linear polynomial $kn + \ell$ assumes prime values; and we may, of course, ask these same questions in relation to any integer valued polynomial (not excluded by trivial considerations). For example, is it true that $n^2 + 1$ is prime for infinitely many values of n? We believe this must happen quite frequently, but the best we can do at present is to show that, infinitely often, $n^2 + 1$ is a number having at most *three* prime factors. There are corresponding results and conjectures for polynomials of higher degree; but for these we refer the reader to Halberstam and Richert, *Sieve Methods* (Academic Press, 1975).

We may ask also whether $kp + \ell$ (HCF$(k,\ell) = 1$, $2|k\ell$) is prime for infinitely many primes p; here $k = 1$, $\ell = 2$ gives the prime twins problem raised in the previous section, and $k = 2$, $\ell = 1$ the problem mentioned in Section 2. Again, we do not know. In 1974 Chen proved that $kp + \ell$ is infinitely often a number having at most *two* prime factors, so that, apparently, we are quite close. However, the gulf separating "two" and "one" is here gigantic.

A closely related question is the famous Goldbach conjecture, now at least 200 years old: is every even number > 4 the sum of two (odd) primes? From Chen's recent work we know that every sufficiently large even number can be written as $p_1 + p_2$ or as $p_1 + p_2 p_3$ in many ways; but as has already been said, removing the second possibility presents formidable difficulties.

The reason that Goldbach's conjecture is so intractable is that it is a "binary" problem — it asks for representation of even n as the sum of *two* numbers of a certain kind. For odd numbers the corresponding problem asks: is every odd integer greater than 2 the sum of at most *three* primes? Here we know from the work of I. M. Vinogradov that this is true for all sufficiently large odd numbers.[i] I have used here, and several times before, this phrase "for all sufficiently large integers". The point is that many analytic methods begin to be effective only for large values, and of course it is the large values that, in principle, give the greatest difficulty because we cannot reach them by direct enumeration. Nevertheless, one does hope eventually to get rid of this limiting phrase; and in this particular instance of representation of numbers as sums of primes, readers may like to know that R. C. Vaughan has recently shown that *all* integers > 1 are sums of at most 27 primes! Thus the primes may be viewed also as an efficient means of building and positive integers > 1 *additively*.

6. There are other integer sequences from which all integers may be constructed by addition of a bounded number of terms. If $\mathcal{B}$ is an integer sequence and every positive integer > 1 is the sum of at most γ elements of $\mathcal{B}$, we say that $\mathcal{B}$ is a *basis of order* γ — of exact order g, if g is the least possible γ for which this is true. Thus the primes form a basis of order 27, and if the full Goldbach conjecture is true then the exact order of the set of primes is 3. In 1770 Lagrange proved, building on the work of Euler, that the natural squares constitute a basis of exact order 4; and in the same year Waring proposed also that, for each $k = 3,4,5, \ldots$, the sequence

$$\mathcal{B}_k : 1^k, 2^k, 3^k, \ldots$$

is a basis. Hilbert was the first to prove, in 1909, the so-called Waring's problem, but the value he obtained for the order $\gamma(k)$ of $\mathcal{B}_k$ was very large. In the years that followed Hardy and Littlewood (building on the work of Hardy and Ramanujan) developed their powerful analytic "circle" method to tackle the problem more efficiently, and later still I. M. Vinogradov sharpened this method in several important respects — his celebrated "three primes" theorem (see Section 5) was one of the triumphs of his improvements. The ramifications of the work on Waring's problem are extensive, and the book of Hardy and Wright — *The Theory of Numbers* (Clarendon, Oxford, 4th ed., 1960) — gives an excellent exposition in chapters XX and XXI. Here we shall just give the gist of the current state of play. Because of computational difficulties, the values of $g(k)$, the exact order of $\mathcal{B}_k$, are still largely unknown — the only precise result we have still is that $g(2) = 4$. The main thrust of endeavour has been towards determining $G(k)$, the least γ such that $\mathcal{B}_k$ is a basis of order γ for all sufficiently large integers. But here too, the only precise result we have, due to Davenport, is that $G(4) = 16$. We know that $G(3) \leqslant 7$ but probably $G(3) = 4$ or 5 so that there is still a large gap to close. Hardy and Littlewood conjectured that

$$G(k) \begin{cases} \leqslant 2k + 1 \ (k \neq 2^m \text{ with } m > 1) \\ \\ = 4k \qquad (k = 2^m \text{ with } m > 1) \end{cases}$$

but so far this has been confirmed only for $k = 3$ and 4. The best we have, for large k, is Vinogradov's result that $G(k)$ is no larger than about $2k \log k$. To improve substantially on Vinogradov's estimate — to prove, for example, that there exists a positive constant C such that $G(k) \leqslant Ck$ — awaits some new ideas, which would need to be of such power as certainly to lead to progress in various other directions too. After half a century of intensive struggle, work in this fascinating area is all but at a standstill.

7. The questions we discussed about the occurrence of primes among the integers, in arithmetic progressions and polynomial sequences generally, may be formulated also for many number sequences other than the primes. An exhaustive account is impossible, but we single out two cases of special interest by way of illustration: (i) the sequence of *s*-free numbers, i.e. integers not divisible by the *s*th power of a prime (the so-called "square-free" integers correspond to $s = 2$); (ii) the sequence of numbers that are sums of two squares. The *k*-free numbers are in plentiful supply — a positive proportion of all integers are *s*-free (for example, about $6/\pi^2$ of all integers are square-free) — and we should therefore expect all the questions to be much easier here. They are not. We know that the least *s*-free number in an arithmetic progression mod *k* occurs before about $k^{1+(1/s)}$, but this falls well short of the expected truth, a constant multiple of *k*. We know that there is always a *k*-free number between *n* and $n + n^{1/(2k)}$ (if *n* is large enough), but again, this falls far short of what is likely to be true. We know that an integer-valued quadratic polynomial takes infinitely many square-free values; but we should expect the same to be true for any integer-valued polynomial (not excluded by obvious considerations), and we are not able to prove this yet. We know only (from a recent result of M. Nair) that if *g* is the degree of such a polynomial, then infinitely often it is $[\lambda g]$-free, where $\lambda = \sqrt{2} - \tfrac{1}{2}$, provided *g* is large enough (for example, it is $(g\text{-}2)$-free if $g \geqslant 18$). Turning to numbers that are sums of two squares, these are only a little more common than the primes (there are about $cx/(\log x)^{1/2}$ of them between 2 and *x*), and very little indeed is known about their finer distribution. For example, it is almost trivial to see that there is always such a number between *n* and $n + n^{1/4}$, yet literally nothing better is known; and the state of ignorance with respect to all the other kinds of questions is no less profound.

8. In paragraph 2 we mentioned inequalities of linear forms in logarithms of primes and several important applications of these. This deep and powerful method relates to a much more general situation and was conceived originally in connection with the problem of identifying classes of *transcendental numbers*, i.e. numbers that are *not* solutions of algebraic equations with integer coefficients.[j] We know that, in a certain precise sense, most numbers are transcendental, yet the problems of deciding whether a given number is transcendental is profoundly difficult.

For example, it is known that *e* and π are transcendental and that so is e^π; but we do not know how to show the same for $\pi + e$ or πe or π^e. Euler's constant γ, mentioned in paragraph 2, is given by

$$\gamma = \lim_{n \to \infty} \left(1 + \frac{1}{2} + \ldots + \frac{1}{n} - \log n \right),$$

and arises in many mathematical contexts. There is an old story that G. H. Hardy declared himself prepared to resign his Chair in Cambridge in favour of anyone who proved that γ was irrational, let alone transcendental! We know that $2^{\sqrt{2}}$ is transcendental and, more generally, that α^{β} is transcendental if $\alpha(\neq 0,1)$ is algebraic and β is algebraic and irrational (proved by Gelford, Schneider independently of each other in 1934); and Baker has shown with his method that the same is true of $\alpha_1^{\beta_1} \ldots \alpha^{\beta}r$ (with the α's, β's similarly constrained provided that there is no linear relation

$$n_0 + n_1\beta_1 + \ldots + n_r\beta_r = 0 \quad (n_i\text{'s integers not all } 0)$$

connecting the β's). However, if β is transcendental, classification of α^{β} is much harder; for example,

$$\beta = \frac{\log 3}{\log 2}$$

is transcendental yet $2^{\beta}(=3)$ is clearly an integer. In this direction we can say that if β is transcendental, at least one of

$$2^{\beta}, 3^{\beta} \text{ and } 5^{\beta}$$

is transcendental — so that, in particular, one of

$$3^{\log 3/\log 2}, \quad 5^{\log 3/\log 2} \tag{8.1}$$

is transcendental — and more general results along these lines can be formulated. Of course, we suspect that each of the numbers (8.1) is transcendental, but this we cannot prove at present.

For this fundamental area of work, the classification of numbers, where the achievements to date have required some of the most difficult and complicated methods in the whole of mathematics, we see clearly the abiding fascination of numbers; nowhere is their opacity more manifest — or more challenging.

Perhaps the reader will understand now what Kronecker had in mind when he compared those mathematicians who concern themselves with number theory to Lotus-eaters who "once having consumed this food can never give it up".

REFERENCES

1. P. Erdös, "Some unsolved problems", *Publ. Inst. Hung. Acad. Sci.* 6, 221-59 (1961).
2. P. Erdös, "Some recent advances and current problems in Number Theory", *Lectures on Modern Mathematics*, Vol. III, edited by T. L. Saaty, Wiley, 1965.
3. D. Shanks, *Solved and Unsolved Problems in Number Theory*, Spartan Books, Washington D.C. 1962.
4. W. Sierpinski, *A Selection of Problems in the Theory of Numbers*, Pergamon Press, 1964.
5. L. J. Mordell, *Diophantine Equations*, Academic Press, New York, 1969.

NOTES

(a) See also L. E. Dickson's *History of the Theory of Numbers* (3 vols.) and the six volumes of *Reviews in the Theory of Numbers* (ed. W. J. leVeque), published by the American Mathematical Society.

(b) p, with or without suffices, always denotes a prime; but $p_1, p_2, \ldots$ denote all the primes written in ascending order only where explicitly stated, as above in (2.1).

(c) See, for example, A. Baker, *Transcendental Number Theory*, Cambridge, 1975.

(d) γ denotes Euler's constant (see paragraph 8). $F(x) = O(x^\delta)$ means that $F(x) \leqslant Ax^\delta$ for some positive constant A.

(e) Primes of the form M_α are known as Mersenne primes.

(f) Perhaps it is of interest to report that quite recently K. Inkeri and A. J. van der Poorten, also M. Nair, have shown that, given a natural number k, all solutions of (3.5) in positive integers x, z and odd primes p, subject to H.C.F. $(x,k) = 1$ and $y - x = k$, are bounded by effectively computable constants depending only on k. The method of proof is, essentially, that used in the proof of Catalan's conjecture (see section 2, p. 193).

(g) A. Selberg proved in 1943, that, if the Riemann Hypothesis is true, then
$$\sum_{n=1}^{N} (p_{n+1} - p_n)^2 = O(N \log^2 N).$$

Recently D. R. Heath-Brown has proved, unconditionally, that the sum on the left is $O(N^{4/3} \log 10^4 N)$. The Riemann Hypothesis, one of the best known conjectures in the whole of mathematics, asserts that the function $\zeta(s)$, defined for $\mathrm{Re}\, s > 1$ as $\sum_{n=1}^{\infty} n^{-s}$ and defined in the rest of the complex plane by analytic continuation, has all its zeros (in the half-plane $\mathrm{Re}\, s > 0$ located on the line $\mathrm{Re}\, s = \frac{1}{2}$.

Apropos of the conjectured existence of primes between m^2 and $m^2 + m$, we know from recent work of Chen that there exist between these numbers many integers each of which is a prime or a product of two primes, provided only that m is large enough.

(h) Euler's function $\phi(k)$ is the number of integers from 1 to k coprime with k, so that there are exactly $\phi(k)$ progressions ℓ mod k with $1 \leqslant \ell < k$ and H C F $(\ell,k) = 1$. If $k = p_1^{\alpha_1} p_2^{\alpha_2} \ldots$, then $\phi(k) = k(1 - 1/p_1)(1 - 1/p_2) \ldots$.

(i) This can be shown to imply that integers from some point onwards are sums of one, two, three or four primes.

(j) Thus all rationals, and all surds, and all numbers derived from these by the elementary operations of addition, subtraction, multiplication and division, are algebraic, i.e. solutions of algebraic equations.

INDEX